T0336991

TEST PROCEDURES FOR SHORT TERM THERMAL STORES

Commission of the European Communities

TEST PROCEDURES FOR SHORT TERM THERMAL STORES

Edited by

H. VISSER

and

H. A. L. VAN DIJK

TNO Institute of Applied Physics,
Delft, The Netherlands

Kluwer Academic Publishers

Dordrecht / Boston / London

Library of Congress Cataloging-in-Publication Data

```
Test procedures for short term thermal stores / edited by H. Visser,
  H.A.L. van Dijk.
       p.   cm.
  At head of title: Commission of the European Communities.
  ISBN 0-7923-1131-0 (HB : acid-free paper)
  1. Solar energy. 2. Heat storage devices--Testing.  I. Visser,
H.  II. Dijk, H. A. L. van.  III. Commission of the European
Communities.
TJ810.T46  1991
621.47'1--dc20                                          91-136
```

ISBN 0–7923–1131–0

Publication arrangements by
Commission of the European Communities
Directorate-General Telecommunications, Information Industries and Innovation, Scientific and
Technical Communications Service, Luxembourg

EUR 13118
© 1991 ECSC, EEC, EAEC, Brussels and Luxembourg

Published by Kluwer Academic Publishers,
P.O. Box 17, 3300 AA Dordrecht, The Netherlands.

Kluwer Academic Publishers incorporates the publishing programmes of
D. Reidel, Martinus Nijhoff, Dr W. Junk and MTP Press.

Sold and distributed in the U.S.A. and Canada
by Kluwer Academic Publishers,
101 Philip Drive, Norwell, MA 02061, U.S.A.

In all other countries, sold and distributed
by Kluwer Academic Publishers Group,
P.O. Box 322, 3300 AH Dordrecht, The Netherlands.

Printed on acid-free paper

Printed in the Netherlands

Contents

Preface

The two most important types of components of active solar heating systems
are, apart from the control system, undoubtedly the solar collector and the
heat storage vessel. Earlier European R&D-work in the frame of the
Commissions Solar Energy R&D-programme has already resulted in European
guidelines for collector test methods *). The present book provides the
test procedures for short term heat stores developed by the European Solar
Storage Testing Group (SSTG).
The kind of research work from which the present book has evolved, is often
referred to as "pre-normative" research, meaning that this type of research
is essential for European harmonization of norms and standards, which in
turn is an essential prerequisite for the Single European Market.
In case of the test methods for solar collectors, the guidelines have
already resulted in agreement on a common European test method in the
UEA-t.c., a European union for agreement on standards in the construction
industry. It can be expected that the proposed test procedures for
characterizing heat storage vessels presented in the current book will
serve a similar purpose. Apart from this the research work has a spin-off
which can be summarized as an improved understanding of the performance of
the heat store and its influence on the total-system performance.
The past ten years have been difficult for researchers in the field of
active solar energy. The relatively low fuel prices reduced the general
interest in their work. Active solar heating systems were generally not
cost-effective. Only solar water heaters in the mediterranean Member States
sold well.
Unless all signs fail, an upturn in the interest in and application of
active solar heating systems seems forthcoming. This upturn is carried by
the waves of a more stable and steady environmental current than the energy
price issue was in the seventies and the eighties.

---

*) "Solar Collectors - Test Methods and Design Guidelines",
edited by W.B. Gillett and J.E. Moon, published for the Commission of the
European Communities under EUR 9778 at D. Reidel Publ. Comp., 1985.

It is thanks to the excellent collaboration in the Solar Storage Testing
Group under the competent guidance of the coordinator Ir. H.A.L. van Dijk
that Europe is a bit better prepared for a renewed interest in active solar
heating systems.

Theo C. Steemers
Commission of the European
Communities

Acknowledgements

The SSTG Co-ordinator would like to acknowledge the enthusiasm, creativity
and team spirit of all members (past and present) of the Commission of the
European Communities' Solar Storage Testing Group (CEC SSTG). This has made
it possible for the Group not only to develop a concise set of proven test
methods for determining the basic thermal properties of a store but also to
carry out the challenging task of creating a number of procedures for
establishing its more detailed characteristics.
The incorporation into the test procedures of software packages for
processing and analysing the test data was a particularly exciting
experience and the contributions made to this part of the programme by
R. Kübler and C.W. Li deserve a special mention. These participants had the
difficult job of producing two fool-proof computer programs - the test
evaluation package STEP (which was developed by R. Kübler) and the
numerical storage model 4PORT (the work of C.W. Li). Anyone who has been
involved in projects of this kind will appreciate the dedication and
accuracy they had to bring to their work. Descriptions of these software
packages, written by the two SSTG members, can be found in Appendices C and
D.
The Co-ordinator is greatly indebted to his colleague, Mr H. Visser, who
brought together all the information produced by the Group into the
coherent test descriptions found in this document. Thanks are also due to
Miss J.R. Stammers for re-editing the text with respect to the English
language. Finally, we must express our sincere appreciation for the
enthusiasm which Mrs A. Helder and Miss H. Windmeijer brought to the task
of preparing the typescript.
The CEC SSTG was initiated by the CEC Directorate-General for Science,
Research and Development (DG XII) as part of the European Communities'
Non-Nuclear Energy R & D Programme. The Co-ordinator would like to put on
record his great appreciation of the stimulating role the CEC's Programme
Manager, Mr T.C. Steemers, played in the activities of the Group. Financial
support came from the CEC and from national agencies of the participating
countries - Denmark, France, the Federal Republic of Germany,
The Netherlands and the United Kingdom.

The present participants have benefited greatly from the work carried out
by the SSTG during its first contractual period, 1981-1983. The first
leader of the Group was my friend and colleague Mr E. van Galen. His
leadership was brought to a cruel end by his sudden illness and death. The
report of the first phase work, the CEC publication 'Recommendations for
European Solar Storage Test Methods', is a fitting memorial to his fine
work, dedication and enthusiasm.

Dick van Dijk
SSTG Co-ordinator

General introduction

This publication is one of the end-products of work carried out during its second contractual period, 1986-1989, by the European Solar Storage Testing Group (SSTG) established by the Commission of the European Communities. The aim of the SSTG was to investigate and develop procedures for the testing of short term solar heat stores. The recommended test procedures resulting from the SSTG programme were primarily developed for the thermal characterization of commercially-available sensible heat storage systems. They can, however, also be used as the starting-point for research and development of new designs and latent heat storage devices. All the recommendations are based on the practical experience gained by the participants during the SSTG concerted actions.

The SSTG participants were:

Denmark   S. Furbo   Thermal Insulation Laboratory
                                   Technical University of Denmark
                                   Building 118
                                   DK-2800 LYNGBY

Federal Republic   J. Sohns   Institut für Thermodynamik und
of Germany   R. Kübler   Wärmetechnik
                                   P.O. Box 801140
                                   D-7000 STUTTGART 30

France   P. Achard   Ecole des Mines
                   E. Nogaret   Centre d'Energétique
                   P. Houzelot   Sophia Antipolis
                   H. Thomas   F-06560 VALBONNE CEDEX

France   C. Buscarlet   Centre Scientifique et Technique du
                                   Bâtiment
                                   Route des Lucioles, BP 141
                                   Sophia Antipolis
                                   F-06561 VALBONNE CEDEX

Netherlands   H.A.L. van Dijk   TNO Institute of Applied Physics
                     (Co-ordinator)   P.O. Box 155
                     J. v.d. Linden   NL-2600 AD DELFT
                     P. Kratz
                     H. Visser

United Kingdom   R.H. Marshall   University of Wales College of Cardiff
                     C.W. Li   Division of Mechanical Engineering
                                          and Energy Studies
                                          Newport Road
                                          UK-CARDIFF CF2 1TA

Observing participants were:

Denmark   T. Vest Hansen   Danish Solar Energy Testing
                     M. Simonson   Laboratory
                     O. Ravn   P.O. Box 141
                                          DK-2630 TAASTRUP

The CEC Programme Manager was:

                     T.C. Steemers   Commission of the European Communities
                                          Directorate-General XII
                                          200 Rue de la Loi
                                          B-1049 BRUSSELS

The complete results of the SSTG programme are presented in two documents. The fundamental work is given in the present publication. It is the result of the Group's co-ordinated activities plus some special subtasks and consists of a note on the SSTG followed by a description of the test facility requirements (Part A) and an outline of the recommended test procedures for heat storage system testing (Part B).

The other document 'Final Report on the Activities of the Solar Storage Testing Group' (EUR 13119) describes the rationale behind the test procedures. It contains an overview of the Group's co-ordinated activities and papers dealing with a number of special participation activities.

The authors should be contacted for information about the availability of the software tools mentioned in this document.

THE CEC SOLAR STORAGE TESTING GROUP

Objectives and scope

The European Solar Storage Testing Group (SSTG) was established at the
beginning of 1982 as a concerted action within the second four-year
(1979-1983) Energy Research and Development Programme of the Commission of
the European Communities. The aim of the Group, which was co-ordinated by
the TNO Institute of Applied Physics (TPD), was to draw up recommendations
for solar storage test procedures. The contract ended in 1983 with the
preparation of the CEC publication 'Recommendations for European Solar
Storage Test Methods' (EUR 9620), the result of one and a half years of
co-ordinated effort by participants from six European countries.
Within the new EC Non-Nuclear Energy R & D Programme (1985-1988), follow-up
concerted actions on storage testing were begun in the course of 1986 to
investigate a number of topics in more detail. Again, the work was
co-ordinated by TPD.

The aim of the work was to develop a set of test procedures for short term
thermal storage systems. Emphasis was given to stores suitable for use in
solar energy domestic hot water and/or space heating systems, i.e. those
with storage volumes of up to 2 m³ working in the temperature range
10-80°C. Currently, commercial storage systems often use water as the
storage medium and due attention was given to such systems in the
programme. So-called latent heat storage systems were also taken into
consideration on the grounds that such systems might be commercialized in
the near future.
The aim was to develop procedures for:
- a simple test method for basic characterization of storage systems, and
- an extensive test method for the stores' detailed characterization.
The latter had to produce parameter values which could be input into a
numerical model of the store under test.

Main types of store and their storage characteristics

An inventory of short term storage systems on the market, carried out as
one the activities of the SSTG, showed that a large proportion of the
commercially-available stores consist of vertical cylinders using water as
the storage material. Usually, the store is charged by means of a helical
coil near the bottom of the store and discharged directly. The most common
heat transfer medium is water. Some commercial storage systems use a mantle
heat exchanger for charge (see Figure 1). A significant number of stores
have either an electric heating element or a built-in fluid heat exchanger
for auxiliary heating of the top of the store. In a few systems the storage
medium is a phase change material. In others, air or water/glycol mix is
the heat transfer fluid. The main field of application of all these stores
is in (pre-)heating domestic hot water and, to a lesser extent, space
heating.

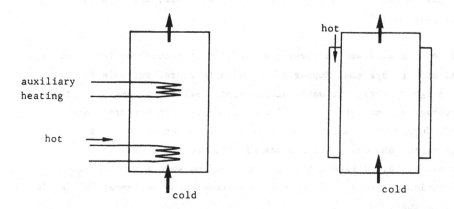

Figure 1: Two types of thermal storage system.

The thermal performance of a store is determined by a number of
characteristics. The main thermal characteristics are:
- storage capacity;
- storage heat loss rate;
- effectiveness of the heat exchanger;
- ability of the store to charge, discharge and/or store the heat in a
  thermally stratified way, rather than with thermal mixing of the storage
  material.

The effect of the capacity on performance is easy to understand: the capacity determines the temperature level and the maximum heat content of the store. The effect of the heat loss rate is equally obvious: it determines the rate at which the heat content of the store leaks to the ambient environment.

The effectiveness of the heat exchanger and the potential to store the heat in a stratified way both affect the time the store will take to charge or discharge. To put it another way, they both influence the 'storage efficiency', i.e. the amount of heat charged/discharged within a certain time compared with the amount of heat intended to be charged/discharged under the same conditions. A partly filled store where the heat is stored in a stratified way can deliver heat at a higher temperature than can a mixed store. This may or may not (according to the application) have a significant effect on the long-term performance of a system. Experience with most of the systems on the market so far has led us to categorize the rate of thermal mixing not as a primary storage characteristic but as a secondary one, i.e. one which is only relevant for detailed characterization or for special designs or applications.

Other examples of secondary characteristics of storage systems are:
- heat loss distribution over the storage volume;
- effectiveness of a built-in heat exchanger for auxiliary heating;
- effectiveness of the heat exchanger at higher or lower flow rates or temperatures;
- thermal capacity at different temperature intervals (in the case of a latent heat store).

The SSTG aimed to investigate how important these characteristics are in the estimation of system performance. To do this, the characteristics have to be translated into the parameters of a numerical model of the store. By integration of the model of the store into a model of the complete heating system, the effect of the store parameters on the (e.g. yearly) system performance can be established. The development of an appropriate storage model and translation of the test results into model parameters was a second major task of the SSTG.

Main types of test and interaction with the storage model

The types of test selected as candidates for the test procedures are
laboratory tests requiring a test facility capable of a defined level of
control and measurement accuracy. Measurements were confined to the
inlet/outlet ports of the store to avoid problems related to the use of
sensors inside the store. The basic procedure is that heat is supplied to
or extracted from the store using a fluid at a certain flow rate and
temperature. The control of the flow rate and temperature depends on the
aim of the test. An example is shown in Figure 2. Here, a step change in
temperature of the fluid entering the store is followed by a change in
temperature of the fluid at the exit. The shape of the curve contains
information on the store characteristics.

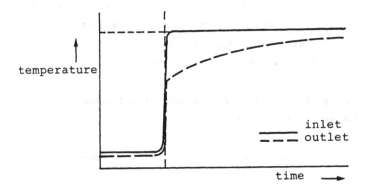

Figure 2: Illustration of a temperature step response test.

In the development of the test procedures the main task was to select
appropriate tests and produce clear instructions on when and how the tests
should be executed.

Some parameters cannot be deduced by direct processing of the experimental data and part of the SSTG's effort was devoted to developing a parameter identification method capable of extracting the needed information in an indirect way. In this method the practical test is simulated using the numerical model of the store and the calculated exit temperature is compared with the measured one.

The unknown values of the parameters are guessed in the first instance and then progressively corrected using an iterative procedure until the best fit is achieved (see Figure 3).

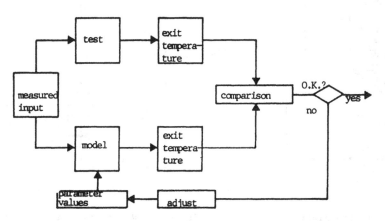

Figure 3: The principle of parameter identification.

The above discussion of the parameter identification technique implies immediately that a numerical model of the store is needed. Figure 4 illustrates how the basic tests, together with information on the store type and the results of parameter identification using the storage model, lead to establishment of the primary characteristics of the store and to a decision on whether additional tests are required. Eventually, the storage model is fed with all the parameters derived from the tests and a dynamic test is run to validate the model. If the model fails, then more detailed tests and/or test evaluations are needed to obtain more refined information.

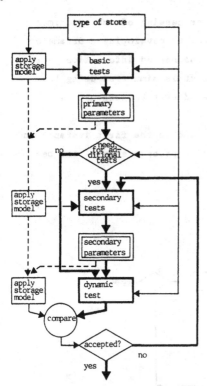

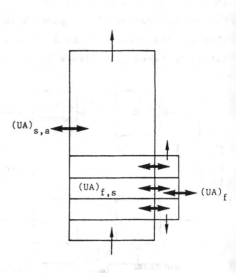

Figure 4: Schematic flow diagram of   **Figure 5**: Numerical model
             the test procedure.

The numerical storage model developed by the SSTG is a finite difference
model in which the store and heat exchanger (if present) are represented by
a number of segments as shown in Figure 5. Each segment has a uniform
temperature. The heat exchange between the segments, or nodes, and between
the nodes and the ambient environment is determined by their temperature
differences and heat exchange coefficients or 'UA'-values - for instance,
the heat transfer from the heat transfer fluid to the storage material and
the heat loss from the store to the ambient environment. The way the store
and heat exchanger are divided depends on, among other things, geometry.
The physical meaning of each segment is thus preserved, which facilitates
the translation from model coefficients to physical characteristics and
vice versa.

The SSTG work plan

The SSTG work plan was designed to allow interaction to take place between
the different parts of the programme - for instance, the selection of
appropriate tests, the development of test procedure details, the
estimation of required levels of accuracy and the development of the
numerical storage model.

The SSTG work plan was divided into joint actions, a special task and a
number of specific subtasks.

The joint action programme was organized round the following series of
tests in which all countries participated:

- Test series A

  A Round Robin test on a selected commercially-available sensible heat
  storage system;

- Test series B

  Tests on difference types of commercially-available heat storage system;

- Test series C

  Special tests.

Through these, the test procedures were developed and improved and the
applicability of the test results to derivation of input parameters for the
numerical storage model was investigated.

The special task consisted of the development of the numerical model for
thermal storage systems. Different possibilities were investigated by teams
in France, the UK and The Netherlands.

Eleven subtasks were defined and divided between the participants. Each
dealt with a specific question related either to the development of the
test procedures or the storage model or to possible extensions of
application areas.

PART A:

TEST FACILITY REQUIREMENTS

Contents of Part A

Nomenclature

symbol	quantity	unit
$\dot{m}$	mass flow rate	kg/s
N	total number of measurements	-
n	number of measurements within 15 minutes	-
$\dot{Q}_c$	heating power supplied to the store	W
$\dot{Q}_{el}$	electric power	W
t	time	s
T_a	ambient air temperature	°C
T_e	store outlet temperature	°C
T_i	store inlet temperature	°C
$T_{i,1}$	store inlet temperature at initial steady state	°C
$T_{i,2}$	store inlet temperature at final steady state	°C
t_1	time interval for steady state heat loss calculation	s
t_0	start time of temperature step change	s
t_{95}	time at which 95% of temperature step change has been completed	s
$\dot{V}$	volume flow rate	m³/s
$\Delta T_{i,e}$	temperature difference between inlet and outlet of the store	K

1. <u>INTRODUCTION</u>

This part (Part A) specifies the minimum requirements of the test facility
for carrying out the thermal characterization test procedures described in
part B. A distinction is made between:
- those requirements which relate to the range and control of the
  adjustable flows and temperatures of the test facility, and
- those requirements which relate to the accuracy of measurement.
Chapter 2 deals with the control of flow rate, store inlet temperature,
ambient air temperature and constant heating power.
The maximum permitted errors in measured flow rate, inlet or outlet
temperature of the store, the temperature difference across the store and
ambient temperature are included in Chapter 3.
Recommendations for test facility design and for meeting the required
accuracy of measurement are outlined in Appendix A.

2. TEST INSTALLATION

2.1. General

The test facility must be designed to meet the test conditions for flow
rate, store inlet temperature, ambient air temperature and heating and/or
cooling power specified in the test descriptions in Part B.
The test facility must be capable of maintaining a flow rate at least in
the range 0.015 10^{-3} m³/s (preferably 0.005 10^{-3} m³/s) to 0.15 10^{-3} m³/s,
i.e. approximately 50 l/h (20 l/h) to 500 l/h.
For the temperature step tests described in Part B Chapter 2, the facility
must be able to heat or cool the store continuously using a heating or
cooling power in the range zero to 10 kW or zero to 5 kW respectively. For
direct discharge tests carried out with an open loop, there is a different
requirement: the facility must be capable of feeding the store once with a
fluid volume 1.2 times its liquid storage volume at a constant temperature
near ambient air temperature.
The test facility has to be able to continuously control the temperature of
the fluid in the ambient to 80°C range.
For the dynamic test (Part B Chapter 5), the facility must be capable of
maintaining the heating power at various constant power levels below
10 kW.

2.2. Flow rate control

All tests require a constant flow rate of the heat transfer fluid. To
achieve this the test facility must meet the following requirements:
- the standard deviation of the flow rate throughout the test (j = 1 → N)
  should not exceed 5% of the mean value:

$$\sigma(\dot{m}) = \left(\sum_{j=1}^{N} \frac{(\dot{m}_j - \bar{m})^2}{N - 1} \right)^{1/2} \leq 0.05 \, \bar{m} \qquad (2.1)$$

where $\dot{m}_j$ is the value of the mass flow rate at the jth measurement,
$\bar{m}$ the arithmetic mean and N the total number of measurements.

- the value of the flow rate integrated over a time interval of 15 minutes
  should at no moment deviate more than ± 5% from the mean flow rate:

$$\left| \frac{1}{n} \sum_{j=k}^{k+n} \dot{m}_j - \bar{\dot{m}} \right| \leq 0.05\ \bar{\dot{m}} \qquad \text{for } k = 1 \rightarrow N - n \qquad (2.2)$$

where n is the number of flow rate measurements carried out within 15
minutes.

2.3. Store inlet temperature control

A number of tests require the store inlet temperature to be constant for
charging and discharging. For these tests the facility must meet the
following requirements:
- the standard deviation of the inlet temperature should not exceed 1 K
  during the period over which a constant temperature is required. Thus:

$$\sigma(T_i) = \left(\sum_{j=1}^{N} \frac{(T_{i,j} - \bar{T}_i)^2}{N - 1} \right)^{1/2} \leq 1\ K \qquad (2.3)$$

where $T_{i,j}$ is the value of the inlet temperature at the jth measurement
and $\bar{T}_i$ the arithmetic mean of all measurements.
- the value of the inlet temperature integrated over a time interval of 15
  minutes should never deviate more than ± 1 K from the mean inlet
  temperature. Therefore:

$$\left| \frac{1}{n} \sum_{j=k}^{k+n} T_{i,j} - \bar{T}_i \right| \leq 1\ K \qquad \text{for } k = 1 \rightarrow N - n \qquad (2.4)$$

For the inlet temperature in the time period for steady state heat loss
calculations (t_1 in Part B Section 2.6.2.), the requirement is more
strict. The mean inlet temperature over the 15 minutes at the beginning of
the heat loss calculations must not differ more than ± 0.1 K from the mean
inlet temperature during the last 15 minutes of the test.

2.4. Quality of inlet temperature step

Various tests require a positive or negative temperature step change in the
inlet temperature of the store.

The step change is assumed to start at t_0, the moment at which the inlet
temperature change exceeds the level of insignificant fluctuations found in
a steady state situation. A suitable criterion for this is a change of
± 1.5 K/min.

The step change is assumed to end at t_{95}, the first moment at which the
change in inlet temperature $|(T_i(t) - T_{i,1}|$ exceeds the value
$0.95 \cdot |T_{i,2} - T_{i,1}|$, where $T_{i,1}$ and $T_{i,2}$ are the mean inlet temperatures
when the system is in a steady state before and after the step change.
Thus, the determination of t_{95} requires a knowledge of the mean inlet
temperature for the period after the step change.

The test facility must be capable of producing a rapid temperature step
without too much of an overshoot, thus:

$$t_{95} < 300 \text{ s} \tag{2.5}$$

2.5. Ambient air temperature control

All tests require constant ambient air temperature and the environment
where the test facility is mounted must meet the following requirements:
- the standard deviation of the ambient air temperature (i.e. the average
  value of the temperature at all locations close to the store) over the
  period of the test should not exceed 2 K. Thus:

$$\sigma(T_a) = \left(\sum_{j=1}^{N} \frac{(T_{a,j} - \bar{T}_a)^2}{N - 1} \right)^{1/2} \leq 2 \text{ K} \tag{2.6}$$

where $T_{a,j}$ is the value of the ambient air temperature at the jth
measurement and $\bar{T}_a$ is the arithmetic mean of all measurements.

- the value of the ambient temperature at each location near to the store
  integrated over 15 minutes should never deviate more than ± 2 K from the
  mean ambient air temperature (see [1], Part B Chapter 5):

$$\left| \frac{1}{n} \sum_{j=k}^{k+n} T_{a,j} - \overline{T}_a \right| \leq 2 \text{ K} \qquad \text{for } k = 1 \rightarrow N - n \qquad (2.7)$$

2.6. Constant heating power control

The dynamic test requires various constant heating power levels. For this
test the facility must meet the following requirements:
- the standard deviation of the heating power during the time when constant
  power is required should not exceed 10% of the mean value:

$$\sigma(\dot{Q}_c) = \left(\sum_{j=1}^{N} \frac{(\dot{Q}_{c,j} - \overline{\dot{Q}}_c)}{N - 1} \right)^{\frac{1}{2}} \leq 0.10 \, \overline{\dot{Q}}_c \qquad (2.8)$$

where $\dot{Q}_{c,j}$ is the value of the heating power supplied to the store at
the jth measurement and $\overline{\dot{Q}}_c$ is the aritmetic mean of all measurements
concerned.
- the value of the heating power integrated over a time period of 15
  minutes should never deviate more than ± 10% from the mean value of the
  heating power:

$$\left| \frac{1}{n} \sum_{j=k}^{k+n} \dot{Q}_{c,j} - \overline{\dot{Q}}_c \right| \leq 0.10 \, \overline{\dot{Q}}_c \qquad \text{for } k = 1 \rightarrow N - n \qquad (2.9)$$

3. INSTRUMENTATION

3.1. Sensor types

The tests described in Part B for characterization of the heat store and
model validation require the measurement of four or more experimental
variables. The four basic ones are:
- the flow rate of the heat transfer fluid, either the volume flow rate $\dot{V}$
  or the mass flow rate $\dot{m}$;
- the temperature of the heat transfer fluid at the store's inlet (supply
  pipe) - store inlet temperature T_i - and/or at the store's outlet
  (return pipe) - store outlet (exit) temperature T_e;
- the difference between the temperatures of the heat transfer fluid at the
  inlet and the outlet of the store $\Delta T_{i,e}$;
- the ambient air temperature T_a;

For tests which use an electric heater the electric power supplied to the
heater $\dot{Q}_{el}$ has to be measured too.

These experimental variables should be measured using flow rate,
temperature, differential temperature and electric power sensors which are
suitable for the operating ranges required by the tests. Calibration of the
sensors should be carried out over these operating ranges roughly every six
months.

The following Sections give the maximum permitted errors in the five
experimental variables described above. They concern the final output
(logged value) after passage through the chain of instruments which follow
the sensor.

3.2. Flow rate measurement

The flow rate should be measured at the supply side of the store.
The total error in the flow rate of the heat transfer fluid should not
exceed ± 1%.

3.3. Store inlet and outlet temperature measurement

The temperature sensor(s) should be positioned as close as possible to the inlet and/or outlet to the store. Possible systematic errors can be reduced by carrying out offset tests (see Appendix A). Care should be taken to ensure mixing of the fluid at the position of the temperature sensor.
The total error in the inlet and/or outlet temperature of the heat transfer fluid should not exceed ± 0.3 K.

3.4. Store differential temperature measurement

The sensors for measuring the temperature difference of the heat transfer fluid across the store should be positioned according to the requirements given in Section 3.3 for the temperature sensor(s) at the inlet and/or outlet of the store.
The total error in the temperature difference should not exceed ± 0.05 K. This is a low value and special care is required (such as the carrying out of offset tests) to avoid possible systematic errors, particularly those due to the positioning of the probes in the test loop.
Recommendations for improving the accuracy of the temperature differential measurements are given in Appendix A.

3.5. Ambient air temperature measurement

The ambient air temperature should be taken as the arithmetic average temperature of the test environment, determined by four sensors calibrated at about 20°C. The sensors should be located at regular distances in a horizontal plane at the vertical midpoint of the store, approximately 0.6 m from its sides. Each sensor should be shielded from radiation effects. The average error should not exceed ± 0.5 K.

3.6. Electric power measurement

The total error in the electric power supplied to the electric auxiliary
heater should not exceed ± 1%.

3.7. Data recording equipment

A permanent record of the variables measured during the tests should be
made by connecting the transducers to a data logging system. The data
logging system should be calibrated regularly and, where possible, the
sensors and data recording equipment should be calibrated together. An
independent cross-check of the calorimetric measurement using a reference
heater may be useful (see Appendix A).

The data recording equipment should be capable of measuring 10 s and 1
minute instantaneous values. The sampling of data at 5 minute intervals is
recommended in steady state periods of the tests.

The resolution of the data logging system should be such that differences
of the order of one fifth of the maximum permitted errors for each variable
being measured can be observed.

TEST PROCEDURES FOR THERMAL CHARACTERIZATION OF STORAGE SYSTEMS

Contents of part B

Nomenclature

symbol	quantity	unit
$C_{s,m}$	measured storage heat capacity	J/K
$C_{s,m,top}$	measured storage heat capacity for top part of the store	J/K
$C_{s,t}$	theoretical storage heat capacity	J/K
c_p	fluid specific heat capacity at constant pressure	J/kgK
dQ_1	relative heat loss of the store	-
E_E	relative error between calculated and measured energy	-
E_T	relative error between calculated and measured outlet temperature of the store	-
f	fraction of storage segment filled with fluid from the downstream segment in one calculation time step	-
h	height in the store	m
L_f	length of the heat exchanger	m
L_s	length of the store	m
$\dot{m}$	mass flow rate	kg/s
$\dot{m}c_p$	heat capacity rate of the heat transfer fluid	W/K
N_f	number of fluid segments	-
N_s	number of storage segments	-
$N_{s,min}$	minimum number of storage segments for an adequate simulation of the heat store	-
N_{dead}	number of segments in a 'dead volume' of the store	-
$Q_{d,1}$	energy withdrawn from the store during partial discharge	J
$Q_{1,sb}$	heat loss of the store during stand-by	J
$Q_{1,top}$	heat loss of the upper part of the store	J
$Q_{s,c}$	calculated heat content of the store	J
$Q_{s,m}$	measured heat content of the store	J
$Q_{s,m}^*$	measured heat discharged from the store not corrected for heat loss	J
$\dot{Q}_c$	heating power supplied to the store	W
$\dot{Q}_{1,ff}$	heat loss rate of the store under finite flow conditions	W
$\dot{Q}_s$	net heating power supplied to or net cooling power withdrawn from the store	W
T_a	ambient air temperature	°C

$T_{a,sb}$	mean ambient temperature over the stand-by period	°C
T_e	outlet temperature of the store	°C
$T_{e,c}$	calculated outlet temperature of the store	°C
$T_{e,1}$	outlet temperature of the store at initial steady state	°C
$T_{e,2}$	outlet temperature of the store at final steady state	°C
T_f	heat transfer fluid temperature	°C
T_i	inlet temperature of the store	°C
$T_{i,1}$	inlet temperature of the store at initial steady state	°C
$T_{i,2}$	inlet temperature of the store at final steady state	°C
T_s	temperature of the store	°C
$T_{s,c}$	calculated temperature of the store	°C
$T_{s,1}$	temperature of the store at initial steady state	°C
$T_{s,2}$	temperature of the store at final steady state	°C
t	time	s
t'	time at certain moment during temperature step test	s
t_{cf}	charge fill time, i.e. the time needed to charge the store by a step change in inlet temperature assuming a perfect heat exchanger and a perfectly stratified store	s
t'_{cf}	charge fill time defined in the dynamic test	s
t_d	discharge test time	s
$t_{d,end}$	time of the end of the partial discharge	s
$t_{d,ini}$	time of the start of the partial discharge	s
t_{df}	discharge fill time, i.e. the time needed to discharge the store by a step change in inlet temperature assuming a perfect heat exchanger and a perfectly stratified store	s
t_1	time interval for steady state heat loss calculation	s
t_{sb}	stand-by time interval	s
$t_{sb,end}$	end of the stand-by time interval	s
t_{ss}	stabilization time	s
$t_{ss,m}$	minimum stabilization time for a mixed store without heat exchanger	s
$t_{ss,min}$	minimum stabilization time	s
t_0	start time of temperature step change	s
$UA)_{f,a}$	mean overall coefficient of heat transfer between heat transfer fluid and ambient environment both under finite conditions and during stand-by	W/K

$(UA)_{f,a,ff}$ overall coefficient of heat transfer between heat transfer fluid and ambient environment under finite flow conditions   W/K

$(UA)_{f,a,sb}$ overall coefficient of heat transfer between heat transfer fluid and ambient environment during stand-by   W/K

$(UA)_{f,s}$ overall coefficient of heat transfer between heat transfer fluid and storage material   W/K

$(UA)_{f,s}^{model}$ overall coefficient of heat transfer between heat transfer fluid and storage material determined by a mixed segment store model   W/K

$(UA)_{f,s,aux}$ auxiliary heater overall coefficient of heat transfer between heat transfer fluid and storage material   W/K

$(UA)_{s,a}$ mean overall coefficient of heat transfer between storage material and ambient environment both under finite flow conditions and during stand-by   W/K

$(UA)_{s,a,ff}$ overall coefficient of heat transfer between storage material and ambient environment under finite flow conditions   W/K

$(UA)_{s,a,sb}$ overall coefficient of heat transfer between storage material and ambient environment during stand-by   W/K

$(UA)_{s,a,top}$ overall coefficient of heat transfer between the upper part of the store and the ambient environment   W/K

$(UA)_{s,s}^{num}$ overall coefficient of heat transfer between storage segments under finite flow conditions caused by numerical diffusion of mathematical model   W/K

$(UA)_{s,s,dd+sb}$ overall coefficient of heat transfer between storage segments combined for direct discharge and stand-by   W/K

$(UA)_{s,s,dd+sb}^{model}$ overall coefficient of heat transfer between storage segments combined for direct discharge and stand-by determined by a mixed segment store model   W/K

$(UA)_{s,s,ff}$ overall coefficient of heat transfer between storage segments under finite flow conditions   W/K

$(UA)_{s,s,ff}^{model}$ overall coefficient of heat transfer between storage segments under finite flow conditions determined by a mixed segment model   W/K

V_s	storage volume	m^3
x	distance in the flow direction	m
$\Delta T_{i,e}$	temperature difference between inlet and outlet of the store	K
Δt	calculation time step	s
ε	heat exchanger effectiveness, i.e. the ratio of the actual energy flow between the heat transfer fluid and the storage material to the maximum energy flow which would occur with an infinite heat exchanger in the store and a uniform storage material temperature	-
$\eta_{s,c}(t)$	storage efficiency during temperature step charge, i.e. the ratio of the net heat supplied to the store up to time t and the total change in heat content between two steady state conditions	-
$\eta_{s,d}(t)$	storage efficiency during temperature step discharge, i.e. the ratio of the heat extracted from the store up to time t and the total change in heat content between two steady state conditions	-
λ_{eff}	mean effective thermal conductivity both under finite flow conditions and during stand-by	W/mK
$\lambda_{eff,dd}$	effective thermal conductivity within the store under direct discharge conditions	W/mK
$\lambda_{eff,dd+sb}$	mean effective thermal conductivity within the store combined for direct discharge and stand-by	W/mK
$\lambda_{eff,ff}$	effective thermal conductivity within the store under finite flow conditions	W/mK
$\lambda_{eff,sb}$	effective thermal conductivity within the store during stand-by	W/mK
ρ	fluid density	kg/m^3

Non-dimensional quantities:

NTU	number of transfer units of the heat exchanger:
	$NTU = (UA)_{f,s}/\dot{m}c_p$
θ_e	non-dimensional outlet temperature of the store in the temperature step test:
	for charge : $\theta_e = (T_e(t) - T_{i,1})/(T_{i,2} - T_{i,1})$
	for discharge: $\theta_e = (T_e(t) - T_{i,2})/(T_{i,1} - T_{i,2})$
θ_i	non-dimensional inlet temperature of the store in the temperature step test:
	for charge : $\theta_i = (T_i(t) - T_{i,1})/(T_{i,2} - T_{i,1})$
	for discharge: $\theta_i = (T_i(t) - T_{i,2})/(T_{i,1} - T_{i,2})$
τ	non-dimensional time in the temperature step test:
	for charge : $\tau = t/t_{cf}$
	for discharge: $\tau = t/t_{df}$

1. INTRODUCTION

1.1. Quantities for heat storage characterization

The tests and combinations of tests described in this part (Part B) produce
values for the primary and secondary thermal properties of a store.
The primary characteristics are:
- the heat storage capacity;
- the heat storage efficiency, i.e. the ratio of net heat supplied to the
  store or heat extracted from it up to time t to the total change in heat
  content between two steady state conditions;
- the heat loss of the store under finite flow conditions;
- the overall coefficient of heat transfer of the heat exchanger, if one is
  present.
The secondary characteristics needed for a more detailed assessment of the
store are:
- the effective thermal conductivity within the store under finite flow
  conditions and during stand-by. These represent the effect of thermal
  conductivity in the storage and tank materials and possible heat exchange
  by the mixing of the fluid, enhanced by the impulse of the fluid entering
  the store, the movement of the fluid itself and (in case of a negative
  vertical temperature gradient) by natural convection. Thus, they indicate
  the store's potential for thermal stratification.
- the overall coefficient of heat transfer of an auxiliary heater for
  fluid, if one is present;
- the heat loss of the store during stand-by;
- the heat loss distribution over the store surface;
- the dependency of the heat exchanger on flow rate and transfer fluid and
  store temperatures.
It will not always be necessary for the secondary characteristics of a
store to be established. Default values for these quantities have therefore
been provided to allow an accurate prediction of performance to be made in
such a case. The default value of the effective thermal conductivity λ_{eff}
is either zero or infinite depending on the temperature gradient in the
store. Thus:

$\quad$ if $dT_s/dh > 0$, then $\lambda_{eff} = 0$ $\qquad\qquad\qquad\qquad$ (1.1)

$\quad$ (temperature increases with increasing height)

if $dT_s/dh < 0$, then $\lambda_{eff} = \infty$ (1.2)

(temperature decreases with increasing height)

where T_s is the store temperature and h is the height in the store.
In the first case, the thermal stratification in the store remains
unchanged; in the second case, cold regions within the store are
immediately mixed with hot regions below them.
For the default situation, the storage heat loss during stand-by is equal
to the loss under flow conditions and the heat loss is equally distributed
over the store surface of the store. The default overall coefficient of
heat transfer of the heat exchanger is a constant value.
It should be noted that the value for storage efficiency takes into account
both the overall coefficient of heat transfer of the heat exchanger and the
rate of thermal mixing during flow.
The set of basic tests to which all heat storage systems should be
subjected plus the additional set of tests for a more detailed
characterization are described in Chapter 7. Here, too, will be found
guidelines to help the user decide whether the detailed characterization is
in fact needed.

1.2. Tests for heat storage characterization

Chapters 2-5 describe the four basic types of test which can be used for
determining store charactistics and validating numerical models of the heat
storage device under test.
For each type of test the objective is given together with a description of
the test method, the requirements for the test rig and instrumentation, the
test conditions, the measurements to be carried out and data analysis for
determining the heat store characteristics or validating the model.
All the test procedures have been developed and used by members of the
SSTG. A distinction is made between:
- those tests which have been tried in different laboratories. These are
  considered to be fully developed.
- those tests which have been tried on a more limited scale. These are
  considered to be provisional and are identified in the text by a
  different type face.

1.3. <u>Processing of the experimental data</u>

The experimental data from the tests are processed partly by analytical
methods and partly by using a parameter identification technique. The
latter involves simulation of the test using an appropriate mathematical
model of the store. The calculated exit temperature is compared with the
measured one and the initially guessed values of the parameters under
consideration are corrected in an iterative way until optimal agreement is
reached.

Computer programs containing both the analytical and the identification
techniques are described briefly in Chapter 6. Again, different type faces
are used to distinguish between fully developed and provisional test
evaluation procedures and programs.

The program for analytical evaluation also calculates the accuracy of the
end results for the main characteristics under examination by determining
the possible measurement errors. An explanation of this is given in
Appendix B.

Recommendations for the content of a heat storage test report are given in
appendix E.

2. TEMPERATURE STEP TEST

2.1. Objective

The temperature step charge or discharge test can be used to determine the following characteristics for all types of store:
- the heat loss of the store under finite flow conditions;
- the heat storage capacity;
- the heat storage efficiency.

Certain other characteristics can be established for particular types of store. For instance, the following can be determined for a store where charge or discharge is carried out by means of a heat exchanger:
- the overall coefficient of heat transfer between the heat transfer fluid and the storage material.

In the case of a store where charge/discharge is carried out using a high performance heat exchanger distributed over the entire store or where there is direct charge/discharge of the storage volume, the following can be determined:
- the effective thermal conductivity within the store.

The quantification of the latter should be regarded as a provisional evaluation procedure.

A modified version of the temperature step test can also be used as a provisional procedure for characterizing a fluid auxiliary heater. The overall coefficient of heat transfer of this heat exchanger can be estimated from a temperature step change on the inlet of the auxiliary heater.

2.2. Description of the test method

In the temperature step test the storage device is charged or discharged from one steady state to another using a temperature step change in the inlet temperature of the store. If the store is charged or discharged by way of a heat exchanger, then the heat transfer fluid is used for this procedure. If the store uses a direct flow mode, then the storage fluid is replaced by a fluid of a different temperature.

The step response test has to begin and end with the storage inlet
temperatures $T_{i,1}$ and $T_{i,2}$ in a steady state, as shown in Figure 2.1
where T_i is the storage inlet temperature, T_e the storage outlet
temperature and t the time. Under these conditions, the heating power
required to maintain steady state is determined. This power is the heat
loss rate of the store under finite flow conditions, $\dot{Q}_{l,ff}$. From the heat
loss rate and the temperature difference between the storage material and
the ambient air, the overall coefficient of heat transfer between the
storage device and the ambient environment under flow conditions
$(UA)_{s,a,ff}$ (in fact $(UA)_{s,a,ff} + (UA)_{f,a,ff}$) is determined.

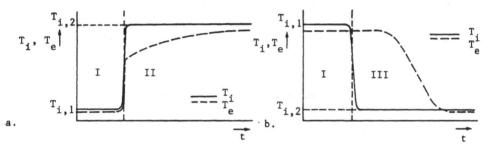

Figure 2.1: Examples of temperature step responses in
   (a)   a charge test using a small heat exchanger and
   (b)   a direct discharge test.
   I.   initial stabilization
   II.   charge step and final stabilization
   III. discharge test and final stabilization

The energy supplied to or withdrawn from the store during the temperature
step is determined. This is the measured charged or discharged heat
$Q_{s,m}$. From this and the temperature interval of the test, the heat
storage capacity $C_{s,m}$ is derived.

The same test data are used to determine the heat storage efficiency,
$\eta_{s,c}$ or $\eta_{s,d}$. This is the ratio of the actual energy supplied to or
withdrawn from the storage device over a certain time to the measured total
amount of heat charged to or discharged from the store in the temperature
interval defined by the temperature step change of the test. The storage
efficiency is determined over various time periods.
The overall coefficient of heat transfer between the heat transfer fluid
and storage material $(UA)_{f,s}$ and the effective thermal conductivity
within the store under finite flow conditions $\lambda_{eff,ff}$ are determined from
the temperature step test data by using a suitable identification model.

The temperature step test for characterization of the fluid auxiliary
heater slightly differs from the test described above. Generally the steady
state situation is only reached after a very long stabilization period.
Consequently, the test is not very suitable for determining the heat loss
and the storage capacity of the part of the store heated by the auxiliary
heater.
The test data are used to estimate the auxiliary heater overall coefficient
of heat transfer $(UA)_{f,s,aux}$. Again, a suitable identification model is
used for the determination.

2.3. Requirements

A test installation is required which can maintain (within the tolerances
given in Part A Chapter 2) the flow rate and the storage inlet temperature
constant at the required conditions during the charge/discharge test time.
It must also be capable of bringing about a well-defined (positive or
negative) temperature step change in the inlet temperature of the store.
The instrumentation required to carry out the measurements and the required
accuracy of the sensors is described in Part A Chapter 3. In general,
temperature (difference) and flow rate sensors should be installed at every
combination of inlet-outlet ports of the store under test.

2.4. Test conditions

2.4.1. Recommended flow rate

In the step response tests the flow rate should be such that the time
required to charge a fully stratified store is about 2 hours. In other
words, the charge fill time $t_{cf} \approx 2$ hrs.
Therefore,

$$\dot{m} = \frac{C_{s,t}}{7200 . c_p} \tag{2.1}$$

unless otherwise recommended by the manufacturer of the store. In equation
(2.1) $C_{s,t}$ is the theoretical storage heat capacity and c_p the specific
heat capacity of the heat transfer fluid.
For relatively large stores the flow rate may be smaller. In these cases
the flow rate is prescribed by the recommended temperature step (see
Section 2.4.2.) and the maximum heating power of the test installation.
Manufacturer's data can be used to calculate $C_{s,t}$.
The discharge flow rate should be approximately 0.10 kg/s (i.e. 6 l/min)
unless otherwise recommended.
For less conventional types of store (e.g. latent heat storage devices) and
for certain specific applications the temperature step change may be
performed at a different or at more than one flow rate.

2.4.2. Recommended temperature step

For sensible heat storage devices the initial steady state temperature of
the charge test should be close to the ambient temperature (i.e. $T_{i,1} \approx T_a$) and the final steady state should be at $T_{i,2} \approx 50°C$. For the
discharge test the initial and final steady state should be $T_{i,1} \approx 50°C$
and $T_{i,2} \approx T_a$ respectively. In direct discharge step tests mains cold
may be used for $T_{i,2}$.

For latent heat storage devices a series of temperature step response tests should be carried out, the storage inlet temperature step change being about 10 K for each step. The series of steps should cover the complete design temperature range. One of the charge and discharge tests should be carried out symmetrically around the transition temperature.

2.4.3. Minimum stabilization time

To estimate the minimum time required to achieve a steady state situation, $t_{ss,min}$, a distinction is made between four types of store: devices with and without a heat exchanger and with a stratified or a mixed temperature step response.

. For a stratified store without a heat exchanger $t_{ss,min}$ is calculated by:

$$t_{ss,min} = 1.5 \cdot t_{cf} \qquad\qquad (2.2)$$

or

$$t_{ss,min} = 1.5 \cdot t_{df} \qquad\qquad (2.3)$$

where t_{cf} and t_{df} are the charge/discharge fill times, i.e. the times needed to charge/discharge a perfectly stratified store with a perfect heat exchanger.
A typical example of such a device is a hot water store with direct discharge.

. For a stratified and a fully mixed store both with a heat exchanger, $t_{ss,min}$ is calculated by:

$$t_{ss,min} = 1.5 \cdot t_{ss,m} \qquad\qquad (2.4)$$

in which $t_{ss,m}$ is:

$$t_{ss,m} = \frac{1}{k_2} \ln \frac{T_{s,1} - k_0}{k_1 - k_0} \tag{2.5}$$

where $T_{s,1}$ is the storage temperature at initial steady state and:

$$k_0 = \frac{\dot{m}c_p \, \varepsilon \, T_{i,2} + (UA)_{s,a,ff} T_a}{\dot{m}c_p \, \varepsilon + (UA)_{s,a,ff}} \quad , \tag{2.6}$$

$$k_1 = \frac{\dot{m}c_p \, \varepsilon \, T_{i,2} + 1.05 \, (UA)_{s,a,ff} T_a}{\dot{m}c_p \, \varepsilon + 1.05 \, (UA)_{s,a,ff}} \tag{2.7}$$

and

$$k_2 = \frac{\dot{m}c_p \, \varepsilon + (UA)_{s,a,ff}}{C_{s,m}} \tag{2.8}$$

In the equations (2.6) to (2.8) ε is the heat exchanger effectiveness:

$$\varepsilon = 1 - e^{-NTU} \tag{2.9}$$

with NTU the number of transfer units:

$$NTU = \frac{(UA)_{f,s}}{\dot{m}c_p} \tag{2.10}$$

where $(\dot{m}c_p)$ is the heat capacity rate of the fluid.

. For a fully mixed store without a heat exchanger $t_{ss,min}$ is calculated
by:

$$t_{ss,min} = t_{ss,m} \qquad\qquad (2.11)$$

After time $t_{ss,m}$ the error in $(UA)_{s,a,ff}$ due to insufficient steady
state is less than 5%.

The temperature step test for characterization of the fluid auxiliary
heater should last for at least 2 hours after the step change.

2.5. Measurements

The following measurements should be taken so that the quantities listed in
Section 2.1 can be determined:
- the inlet temperature of the store;
- the temperature difference of the heat transfer fluid between the outlet
  and the inlet of the store;
- the flow rate of the heat transfer fluid;
- the ambient air temperature.

The logging rate for measurements taken during the temperature step test
should be as follows:
. 1 minute instantaneous up to 5 minutes before the step change;
. 10 s instantaneous from 5 minutes before the step change up to the
  following times:
  - t_{cf} or t_{df} for a temperature step involving a heat exchanger
  and
  - 1.5 t_{cf} or 1.5 t_{df} for a direct charge or discharge case;
. 1 minute instantaneous for the remainder of the test.
If a test requires long stabilization times, it is recommended that
readings are logged instantaneously every 5 minutes after 8 hours have
elapsed following the step change. This will reduce the amount of
experimental data collected.

The start of the temperature step change is assumed to be at t_0, the measuring point before the time at which the inlet temperature gradient is 100 K/hr or more.

2.6. Calculation of the quantities

2.6.1. Estimation of the store temperature

In order to determine $(UA)_{s,a,ff}$ under steady state conditions, the temperature of the store T_s has to be calculated to a high degree of accuracy. The quantity $(UA)_{s,a,ff}$ is also needed for the calculation of the heat losses during the transient between two steady state conditions. As the temperature of a store normally changes rapidly, a larger error in T_s during the transient is acceptable because it will not affect the total heat loss from one steady state to the other (and thus $C_{s,m}$ and $\eta_{s,c}$ or $\eta_{s,d}$ significantly (see Appendix C).

For the calculation of the store temperature, a distinction is made between four types of store: devices with and without a heat exchanger and with a stratified or a mixed temperature step response. The initial steady state is characterized by $T_{i,1}$ and $T_{e,1}$, the final steady state by $T_{i,2}$ and $T_{e,2}$; $\Delta T_{i,e}$ is the temperature difference between inlet and outlet of the store.

. For a strafified store without a heat exchanger, T_s for steady state ($t < t_0$ or $t > t_{cf}$) is estimated from:

$$T_s = \frac{T_i + T_e}{2} \tag{2.12}$$

and during transient ($t_0 < t < t_{cf}$) from:

$$T_s = \frac{T_{i,1} + T_{e,1}}{2} + t_H \cdot \left(\frac{T_{i,2} + T_{e,2}}{2} - \frac{T_{i,1} + T_{e,1}}{2} \right) \tag{2.13}$$

with

$$t_H = \frac{t - t_0}{t_{cf}} \tag{2.14}$$

. For a stratified store with a heat exchanger, T_s is estimated from:

$$T_s = \frac{T_i + T_e}{2} - \frac{\Delta T_{i,e}}{NTU} \tag{2.15}$$

. For a fully mixed store without a heat exchanger, the store temperature is:

$$T_s = T_e \tag{2.16}$$

. For a fully mixed store with a heat exchanger T_s is estimated from:

$$T_s = T_i - \frac{\Delta T_{i,e}}{1 - e^{-NTU}} \tag{2.17}$$

To avoid physically impossible values of the store temperature ($T_s < T_{s,1}$) due to the underestimation of $(UA)_{f,s}$ during the time the heat transfer fluid is in the heat exchanger, in all cases where $T_s < T_{s,1}$ is calculated, $T_s = T_e$ is set in the evaluation.

The calculation of the possible error in T_s is given in Appendix C.

2.6.2. <u>Determination of the overall coefficient of heat transfer between the store and the ambient environment under finite flow conditions</u>

The overall coefficient of heat transfer between the store and the ambient environment under finite flow conditions $(UA)_{s,a,ff}$ is determined over the time period t_1, by the following:

$$(UA)_{s,a,ff} = \frac{1}{t_1} \cdot \int_{t_{ss}}^{t_{ss} + t_1} \frac{\dot{m}c_p \, \Delta T_{i,e}}{T_s - T_a} \, dt \qquad (2.18)$$

where t_{ss} is the time after which the store has reached steady state. It is recommended that $t_1 = 4$ hrs is used for the calculation of the steady state heat loss.

The calculation of the possible error in $(UA)_{s,a,ff}$ is given in Appendix C.

2.6.3. <u>Determination of the heat storage capacity</u>

The rate of change of the heat content of the store is calculated from:

$$\dot{Q}_s = \dot{m}c_p \, \Delta T_{i,e} - (UA)_{s,a,ff} \, (T_s - T_a) \qquad (2.19)$$

The heat charged or discharged up to time t' is obtained by integration:

$$Q_{s,m}(t') = \int_{t_0}^{t'} \dot{Q}_s \, dt \qquad (2.20)$$

The measured storage heat capacity is given by:

$$C_{s,m} = \frac{Q_{s,m}(t_{ss,min})}{T_{s,2} - T_{s,1}} \qquad\qquad (2.21)$$

where $T_{s,2}$ is the storage temperature at the final steady state.

The calculation of the possible errors in $Q_{s,m}$ and $C_{s,m}$ is given in Appendix C.

2.6.4. Determination of the heat storage efficiency

The storage efficiency for the temperature step charge is the ratio of the net heat supplied to the store until time t and the total net heat charged between two steady state conditions:

$$\eta_{s,c}(t) = \frac{Q_{s,m}(t)}{Q_{s,m}(t_{ss})} \qquad\qquad (2.22)$$

The calculation of the possible error in $\eta_{s,c}$ is given in Appendix C.

For temperature step discharge tests the storage efficiency is calculated in a similar way. However, the heat withdrawn from the store is not corrected for the heat loss because it is the heat content of the discharge flow in which we are interested:

$$\eta_{s,d}(t) = \frac{Q_{s,m}^*(t)}{Q_{s,m}(t_{ss})} \qquad\qquad (2.23)$$

with

$$Q_{s,m}^*(t) = \int_{t_0}^{t_0 + t} \dot{m}c_p \left|\Delta T_{i,e}\right| dt \qquad\qquad (2.24)$$

and $Q_{s,m}(t_{ss})$ calculated from equation (2.20).

2.6.5. <u>Determination of the overall coefficient of heat transfer between
the heat transfer fluid and the storage material</u>

If the temperature step change has been carried out using a heat exchanger,
the overall coefficient of heat transfer between heat transfer fluid and
storage material $(UA)_{f,s}$ is determined using a parameter identification
method. In this, the test is simulated with an appropriate mathematical
model of the store. One of the model parameters is the $(UA)_{f,s}$-value,
which is obtained by minimization of the difference between the measured
and calculated values of the storage outlet temperature. The quantities
already determined from the temperature step like $(UA)_{s,a,ff}$ and $C_{s,m}$
serve as known parameters. The quantity $\lambda_{eff,ff}$ is also a known
parameter, being either determined according to the procedure described in
Section 2.6.6 or assigned a default value as in Section 1.1. When there is
a negative temperature gradient in the store, a mixing routine in the
storage model is equivalent to the default $\lambda_{eff,ff} = \infty$.
Furthermore, the minimum number of store segments to enable the
mathematical model to give an adequate simulation of the behaviour of a
thermally stratified store is defined by the following equation:

$$N_{s,min} = 30 \ (1 - f) \tag{2.25}$$

So that there is resolution in the calculated temperature response, the
number $N_{s,min}$ should never be less than 10. In equation (2.25) f is the
fraction of the store segments which is filled with fluid from the
downstream segments in one calculation time step Δt:

$$f = \frac{N_s \cdot \dot{m} \cdot \Delta t}{\rho \ V_s} \tag{2.26}$$

where N_s is the actual number of storage segments in the model, V_s the
storage volume involved in flow and $\dot{m}$ and ρ the mass flow rate and density
of the fluid respectively.

If the fraction f is not equal to 1 there is intrinsic numerical diffusion between the segments in the model. If $N_s \geq N_{s,min}$, the effect of numerical diffusion on the yearly system performance is insignificant. A detailed discussion on numerical diffusion can be found in the final report on the SSTG activities (see [1], Part A Chapter 3).

The fluid temperature in the heat exchanger is assumed to behave as in a plug flow:

$$\frac{dT_f(x)}{dx} = - \frac{NTU}{L_f} (T_f(x) - T_s) \qquad (2.27)$$

where x is the distance in the flow direction and L_f the length of the heat exchanger. The NTU- or $(UA)_{f,s}$-value is implicitly defined by equation (2.27).

The heat exchange between transfer fluid and storage material is dependent on the flow rate, the fluid and the storage temperature:

$$(UA)_{f,s} = f\ (\bar{T}_f\ ,\ (\bar{T}_f - T_s\ )\ ,\ \dot{m}c_p) \qquad (2.28)$$

where $\bar{T}_f$ is the average temperature of the transfer fluid in the heat exchanger.

An empirical relationship between the $(UA)_{f,s}$-value and the flow rate, the fluid and the storage temperature which is valid for a wide range of heat exchangers is based in Kübler et al [2]:

$$(UA)_{f,s} = c_0 \cdot (\dot{m}c_p)^{C_1} \cdot (T_i - T_s)^{C_2} \cdot ((T_i + T_s)/2)^{C_3} \qquad (2.29)$$

with $(\dot{m}c_p)$ expressed in W/K, the temperature in °C and c_0, c_1, c_2 and c_3 coefficients which depend on the type of heat exchanger as indicated in Table 2.1. To increase the reliability, however, it is recommended that $(UA)_{f,s}$ be determined separately for each flow rate. Furthermore, if the researcher has good reason to believe that another equation is more appropriate in a particular case, he should replace equation (2.29) with the better one.

Table 2.1: Coefficients for the calculation of $(UA)_{f,s}$-values for charge
heat exchangers of various types according to equation (2.29).
Source: [2]

type of heat exchanger	c_1	c_2	c_3
finned coil, vertical	0.352	0.227	0.546
finned coil, horizontal	0.341	0.214	0.638
smooth coil, vertical	0.141	0.184	0.451

As a good approximation the $(UA)_{f,s}$-value in a charge test can be
regarded as constant, since in the course of the test the increase in
the mean temperature more or less compensates for the decrease in
temperature difference.

In any case, the identification model gives a $(UA)_{f,s}$-value weighted over
the test time interval considered. It is recommended that the idenfication
of $(UA)_{f,s}$ be based on the measuring points from $t = t_0$ up to
$t = t_{cf}$, as the identification procedure is most sensitive for the first
part of the step, i.e. the part when most of the energy transfer takes
place.

From the identified $(UA)_{f,s}$-value, the constant c_0 can be derived. If,
however, the store model considers each individual heat exchanger
segment to be mixed, the identification procedure gives a $(UA)_{f,s}^{model}$-value
which needs to be converted using the following equation:

$$(UA)_{f,s} = \dot{m}c_p \, N_f \, \ln \left(\frac{(UA)_{f,s}^{model}}{\dot{m}c_p \, N_f} + 1 \right) \qquad (2.30)$$

where N_f is the number of fluid segments in the modelled heat exchanger.

Now, c_0 can be determined from:

$$c_0 = (UA)_{f,s} / \left\{ (\dot{m}c_p)^{C_1} \cdot (T_i - T_s)^{C_2} \cdot ((T_i + T_s)/2)^{C_3} \right\}_{ref} \quad (2.31)$$

where 'ref' means 'reference'. The reference T_i- and T_s-values should
be representative of the part of the step in which the bulk of the energy
transfer takes place. Again, this is between $t = t_0$ and $t = t_{cf}$. Hence:

$$T_{i,ref} = T_{i,2} \quad\quad\quad\quad\quad\quad\quad\quad (2.32)$$

$$T_{s,ref} = \frac{1}{t_{cf}} \int_{t_0}^{t_0 + t_{cf}} T_s (t) dt \quad\quad\quad\quad (2.33)$$

There is no reason to suppose that the $(UA)_{f,s}$-value of a discharge heat
exchanger can be approximated by a constant value.
In that case it is not the $(UA)_{f,s}$-value but the c_0-value in an equation
like (2.29) which has to be identified using an appropriate identification
model.
Information on the available identification method and models is given in
Chapter 6.

2.6.6. <u>Determination of the effective thermal conductivity within the store under finite flow</u>
       <u>conditions</u> (provisional evaluation procedure)

The effective thermal conductivity within the store under finite flow conditions $\lambda_{eff,ff}$
can be determined from a charge or discharge using of a heat exchanger with a very high
$(UA)_{f,s}$-value distributed over the whole store and from a direct charge or discharge
temperature step test. It will not always be necessary to determine the $\lambda_{eff,ff}$-value
because it is a secondary heat storage quantity. The criterion for establishing whether or
not the quantity should be determined is given in Chapter 7.

For the determination of $\lambda_{eff,ff}$, again, a suitable identification model should be used. In the mathematical model $\lambda_{eff,ff}$ is converted into an overall coefficient of heat transfer between discrete storage segments under finite flow conditions:

$$(UA)_{s,s,ff} = \frac{N_s V_s}{L_s^2} \lambda_{eff,ff} \qquad (2.34)$$

with L_s the length of the store in the flow direction.

The idenfication procedure is identical to that described in Section 2.6.5. In the present case, the $(UA)_{f,s}$-value is one of the known parameters.

The mathematical model identifies $(UA)_{s,s,ff}^{model}$. Because there could be numerical diffusion between the segments in the model, the $(UA)_{s,s,ff}$-value should be corrected as follows:

$$(UA)_{s,s,ff} = (UA)_{s,s,ff}^{model} + (UA)_{s,s}^{num} \qquad (2.35)$$

in which $(UA)_{s,s}^{num}$ is caused by numerical diffusion of the mathematical model during flow:

$$(UA)_{s,s}^{num} = 0.5 \, (1 - f) \, \dot{m} c_p \qquad (2.36)$$

with f according to equation (2.26).

2.6.7. Determination of the fluid auxiliary heater overall coefficient of heat transfer (provisional test)

The fluid auxiliary heater overall coefficient of heat transfer $(UA)_{f,s,aux}$ is determined when a temperature step charge test with the auxiliary heater has been carried out. This requires use of a suitable identification model. The procedure is as described in Section 2.6.5. All the other heat storage quantities are known model parameters.

It is recommended that the experimental data collected during the first 2 hours after the step change be used for the identification of $(UA)_{f,s,aux}$.

3. STAND-BY HEAT LOSS TEST

3.1. Objective

From a stand-by heat loss test, the heat loss during a stand-by period of the heat storage device tested can be determined.

3.2. Description of the test method

In the stand-by heat loss test the storage device is initially charged to a steady state situation either by using the heat exchanger or by direct flow. In the steady state situation the flow from the test installation to the store is interrupted for a stand-by period. At the end of the stand-by period the heat storage is either charged to the steady state it was at before the stand-by or discharged to a different steady state. In Figure 3.1 both types of test have been illustrated.

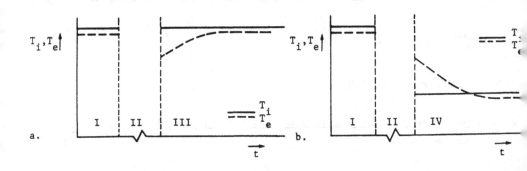

Figure 3.1: Examples of inlet and outlet temperatures of the store in stand-by heat loss tests with (a) a recharge and (b) a discharge after the stand-by period.
I.   initial stabilization;
II.   stand-by;
III. recharge and final stabilization;
IV. discharge and final stabilization.

The heat loss of the store during the stand-by period $Q_{1,sb}$ is determined from the recharged or discharged heat and the difference between initial and final heat content of the store. From the heat loss and the heat content of the store before stand-by, the relative heat loss dQ_1 is obtained. By also using the stand-by time and the heat storage capacity, the overall coefficient of heat transfer between the store and the ambient environment during a stand-by period $(UA)_{s,a,sb}$ (in fact $(UA)_{s,a,sb}$ + $(UA)_{f,a,sb}$) is determined.

3.3. Requirements

The requirements for the test installation and instrumentation are as given in Section 2.3. During stand-by of the test in the reheat mode the heat transfer fluid should by-pass the store in order to maintain the same inlet temperature.

3.4. Test conditions

3.4.1. Recommended flow rate

The flow rate in the charge or discharge parts of the stand-by heat loss test should be the same as that used in the temperature step test. It can be calculated by equation (2.1).

3.4.2. Recommended steady state temperature

The steady state temperature at the beginning of the stand-by period should be the maximum design temperature.
For a latent heat type storage device a second steady state temperature 5 K below the transition range is recommended.

The minimum time required to achieve a steady state situation is the same as that used in the step response tests (see Section 2.4.3).

3.4.3. Recommended stand-by period

In general, a stand-by period of 24 hrs is recommended although for certain specific store applications it may be appropriate to perform the heat loss test with a different stand-by time interval.

3.5. Measurements

The measurements listed in Section 2.5. should be taken, using the following logging rates:
. 5 minute instantaneous for the initial steady state and the stand-by period until 5 minutes before the recharge or discharge;
. 10 s instantaneous from 5 minutes before the recharge or discharge up to the following times:
   - t_{cf} or t_{df} for a temperature step using a heat exchanger and
   - 1.5 t_{cf} or 1.5 t_{df} for a direct charge or discharge case;
. 1 minute instantaneous for the remainder of the test.
If a test requires long stabilization times, it is recommended that readings are logged instantaneously every five minutes after 8 hours have elapsed following the end of the stand-by period.

3.6. Calculation of the overall coefficient of heat transfer between the store and the ambient environment during stand-by

The stand-by heat loss derived from recharging the store is determined by:

$$Q_{1,sb} = \int_{t_{sb,end}}^{t_{sb,end}+t_{ss}} \dot{Q}_s \, dt \qquad (3.1)$$

in which $t_{sb,end}$ is the end of the stand-by period, t_{ss} is the time to reach the initial steady state again and $\dot{Q}_s$ is calculated according to equation (2.19).

When there is a discharge after stand-by the heat loss is calculated from:

$$Q_{1,sb} = C_{s,m}(T_{s,1} - T_s(t_{sb,end} + t_{ss})) - \int_{t_{sb,end}}^{t_{sb,end} + t_{ss}} \dot{Q}_s dt \qquad (3.2)$$

where $T_{s,1}$ is the store temperature before stand-by and $T_s(t_{sb,end} + t_{ss})$ the store temperature after discharge.

The relative heat loss in both cases is determined from:

$$dQ_1 = \frac{Q_{1,sb}}{C_{s,m}(T_{s,1} - T_{a,sb})} \qquad (3.3)$$

where $T_{a,sb}$ is the average value of the ambient temperature over the stand-by period.

Finally, the overall coefficient of heat transfer between the store and the ambient environment during stand-by is calculated from:

$$(UA)_{s,a,sb} = -\frac{C_{s,m}}{t_{sb}} \ln \left[\frac{C_{s,m} \cdot (T_{s,1} - T_{a,sb}) - Q_{1,sb}}{C_{s,m}(T_{s,1} - T_{a,sb})} \right] \qquad (3.4)$$

where t_{sb} is the stand-by time interval.

4. STAND-BY HEAT LOSS TEST AFTER PARTIAL DISCHARGE

4.1. Objective

From a stand-by heat loss test after partial discharge the following
quantities of the heat store can be determined:
- the heat loss of the upper, hot part of the store;
- the effective thermal conductivity within the store during stand-by.
This test is only relevant for sensible heat storage devices which can be
discharged directly.
An indication of the heat loss distribution over the surface of the store
can be obtained by combining the results of one or more partial discharge -
stand-by heat loss tests with the heat loss value of the whole store. To
determine the effective thermal conductivity during stand-by we need to
know the same quantity for direct discharge. Therefore, the results from a
direct discharge step under the same conditions should also be available.
The method of determiningthe effective thermal conductivity during stand-by
is to be regarded as provisional.
The heat loss distribution over the store surface as well as the effective
thermal conductivity during stand-by are secondary characteristics and
therefore do not always have to be determined. The criteria for deciding
whether or not to do so are given in Chapter 7.

4.2. Description of the test method

In the partial discharge - stand-by heat loss test the storage device is
initially charged to a steady state by using the heat exchanger or by
direct flow. In this steady state situation a part of the storage volume is
discharged directly. Instantly after the partial discharge the flow from
the test installation to the store is interrupted for a stand-by period. At
the end of the stand-by period the heat store is either charged to the
steady state it was at before the partial discharge or discharged to a
different steady state. In Figure 4.1 both types of partial discharge -
stand-by heat loss test have been illustrated.

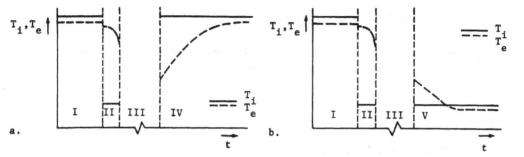

<u>Figure 4.1</u>: Examples of inlet and outlet temperatures of the store for
        partial discharge - stand-by heat loss tests with (a) a
        recharge and (b) a discharge after the stand-by period.
        I.   initial stabilization;
        II.   partial direct discharge;
        III. stand-by;
        IV.   recharge and final stabilization;
        V.   discharge and final stabilization.

The heat loss from the upper part of the store during the stand-by period
$Q_{1,top}$ is determined from the recharged heat and the energy withdrawn
from the store during the partial discharge or from the total discharged
heat and the initial heat content of the store. The overall coefficient of
heat transfer between the upper part of the store and the ambient
environment $(UA)_{s,a,top}$ is determined from the upper part heat loss, the
heat content of the store before stand-by, the stand-by time and the heat
storage capacity.
For the case of a direct discharge after stand-by, the effective thermal
conductivity within the store during stand-by $\lambda_{eff,sb}$ can be determined
from the test data by using a suitable identification model. For the
determination of $\lambda_{eff,sb}$, the same quantity under flow conditions
$\lambda_{eff,ff}$ (see Section 2.6.6) is needed.

4.3. Requirements

The requirements for the test installation and instrumentation are
identical to those given in Section 3.3.

4.4. Test conditions

4.4.1. Recommended flow rate

The flow rate in the charge and/or discharge parts of the test should be
identical to that used in the temperature step tests. It can be calculated
by equation (2.1).

4.4.2. Recommended steady state temperature

The steady state temperature before the partial discharge should be the
maximum design temperature.
The inlet temperature during discharge should be approximately equal to the
ambient temperature.

The minimum time required to achieve a steady state is the same as that
needed for the step response tests (see Section 2.4.3.).

4.4.3. Recommended partial discharge time

It is recommended that about half of the storage volume be withdrawn for
the partial discharge. Therefore, the direct discharge time is:

$$t_d = 0.5 \ t_{df} \qquad\qquad (4.1)$$

If a more detailed heat loss distribution over the surface of the store is
wanted, t_d should have different values.

4.4.4. Recommended stand-by period

In general, a stand-by period between 6 and 24 hrs is recommended although for specific store applications it may be appropriate for it to be different.

4.5. Measurements

The measurements listed in Section 2.5. should be obtained using the following logging rates:

. 5 minute instantaneous for the initial steady state until 5 minutes before the partial discharge;

. 10 s instantaneous from 5 minutes before the partial discharge until the stand-by period;

. 5 minute instantaneous from the beginning of the stand-by period until 5 minutes before the recharge or discharge;

. 10 s instantaneous from 5 minutes before the recharge or discharge until the following times:

  - t_{cf} or t_{df} for a recharge using a heat exchanger

  and

  - 1.5 t_{cf} or 1.5 t_{df} for a direct recharge or discharge case;

. 1 minute instantaneous for the remainder of the test.

If a test requires long stabilization times, it is recommended that readings be logged at a rate of 5 minutes instantaneous from 8 hours after the end of the stand-by period.

4.6. Calculation of the quantities

4.6.1. Determination of the overall coefficient of heat transfer between the upper part of the store and the ambient environment

The stand-by heat loss of the upper part of the store derived from a recharge after stand-by is determined by:

$$Q_{1,top} = \int_{t_{sb,end}}^{t_{sb,end} + t_{ss}} \dot{Q}_s \, dt - Q_{d,1} \qquad (4.2)$$

in which $Q_{d,1}$ is the energy withdrawn from the store during the partial
discharge:

$$Q_{d,1} = \int_{t_{d,ini}}^{t_{d,end}} \dot{Q}_s \, dt \qquad\qquad (4.3)$$

where $t_{d,ini}$ and $t_{d,end}$ are the times of start and end of the partial
discharge respectively and $\dot{Q}_s$ is calculated according to equation (2.19).
For a discharge after stand-by, the heat loss of the upper part of the
store is calculated from:

$$Q_{1,top} = C_{s,m}(T_{s,1} - T_s(t_{sb,end} + t_{ss})) - Q_{d,1} - \int_{t_{sb,end}}^{t_{sb,end} + t_{ss}} \dot{Q}_s \, dt$$

$$(4.4)$$

For both test types the overall coefficient of heat transfer between the
upper part of the store and the ambient environment during stand-by is
calculated from:

$$(UA)_{s,a,top} = -\frac{C_{s,m,top}}{t_{sb}} \ln \frac{C_{s,m}(T_{s,1} - T_{a,sb}) - Q_{1,top} - Q_{d,1}}{C_{s,m}(T_{s,1} - T_{a,sb}) - Q_{d,1}} \quad (4.5)$$

where

$$C_{s,m,top} = C_{s,m} - \frac{Q_{d,1}}{(T_{s,1} - T_a)} \qquad\qquad (4.6)$$

4.6.2. Determination of the effective thermal conductivity within the store during stand-by
(provisional evaluation procedure)

From a partial discharge - stand-by - complete discharge test the mean effective thermal
conductivity within the store under direct discharge and stand-by conditions, $\lambda_{eff,dd+sb}$,
can be determined using a suitable identification model.
In the mathematical model $\lambda_{eff,dd+sb}$ is converted into an overall coefficient of heat

transfer between discrete storage segments under direct discharge and stand-by conditions:

$$(UA)_{ss,dd+sb} = \frac{N_s V_s}{L_s^2} \lambda_{eff,dd+sb} \tag{4.7}$$

The identification procedure is as described in Section 2.6.5. The $(UA)_{f,s}$-value is a parameter in the identification model and is known from a temperature step test (see Section 2.6.5.).

The mathematical model identifies $(UA)_{s,s,dd+sb}^{model}$. The $(UA)_{s,s,dd+sb}$-value should be corrected for possible numerical diffusion between the segments in the model during both discharge periods, using the following equation:

$$(UA)_{s,s,dd+sb} = (UA)_{s,s,dd+sb}^{model} + \frac{t_d}{t_d + t_{sb}} \cdot (UA)_{s,s}^{num} \tag{4.8}$$

in which t_d is the sum of both discharge periods and $(UA)_{s,s}^{num}$ is given by equation (2.36). If a good estimation of the effective thermal conductivity under direct discharge conditions $\lambda_{eff,dd}$ has been obtained from a full direct discharge test (see Section 2.6.6), then the effective thermal conductivity during stand-by can be derived as separate value:

$$\lambda_{eff,sb} = \frac{t_{sb} + t_d}{t_{sb}} \cdot \lambda_{eff,dd+sb} - \frac{t_d}{t_{sb}} \cdot \lambda_{eff,dd} \tag{4.9}$$

with $\lambda_{eff,dd}$ according to equation (2.34).

5. DYNAMIC TEST

5.1. Objective

The objective of the dynamic test is threefold:
- to validate the quantities obtained from test or manufacturer's data for use as parameters in model simulations;
- to investigate the need for additional tests or further processing of the tests already carried out for determining secondary characteristics such as the heat loss distribution over the store surface or the effective thermal conductivity within the store;
- to ascertain the appropriateness of the model configuration for the heat storage device.

To validate the modelling of a built-in fluid auxiliary heater, the dynamic test can also be carried out using the auxiliary heat exchanger to charge the store instead of the normal charge mode. A heat storage model with a built-in electrical auxiliary heater can be validated by carrying out an additional dynamic test under charge, discharge and stand-by using the electric heater as prescribed by the manufacturer for normal operation. The dynamic tests using the fluid or electrical auxiliary heater are regarded as provisional.

5.2. Description of the test method

In the dynamic test the heat storage passes through most (if not all) stages as for normal operation. The store is charged twice under different heating powers and sequences. The test includes a complete discharge as well as a partial discharge followed by a complete discharge after a stand-by period.

The dynamic test described here is suitable for stores which are charged by a principle or auxiliary heat exchanger and discharged directly. An electrical auxiliary heater may or may not be in operation. In general, the complete test sequence covers two days. For special types of store or specific store applications, the dynamic test may be changed or an additional test sequence introduced.

As part of the evaluation, the dynamic test is simulated with a heat storage model using as input parameter values obtained from earlier tests or from manufacturer's data. The results of the simulation are compared with the measurements. The heat storage model is considered to be validated if the comparision meets specific criteria with respect to energy balance and temperature deviation. If these criteria are not met, the storage parameters have to be corrected or a more detailed model obtained which gives a correct numerical description of the store under test. In these cases, the tests already performed have to be processed further or new tests carried out.

5.3. Requirements

A test installation is required which is capable of maintaining the flow rate, the storage inlet temperature and the heating power at the required conditions during the different periods of the dynamic test sequence within the tolerance specified in Part A Chapter 2. The test installation must be capable of bringing about a negative step change in the storage inlet temperature to around the ambient temperature. A constant heating rate at two power levels (full and half power) is required. The particular power levels depend on the storage heat capacity.
The requirements for the instrumentation are as given in Section 2.3.

5.4. Test conditions

5.4.1. Dynamic test sequence

The two days test sequence is especially suited to validating a calculation model with respect to the following parameters:
- the heat storage capacity C_s;
- the overall coefficient of heat transfer between storage device and the ambient environment $(UA)_{s,a}$;
- the overall coefficient of heat transfer between heat transfer fluid and storage material $(UA)_{f,s}$;

- the configuration of the model, in particular the number and location of
  segments in the storage material N_s and the heat transfer fluid N_f.
Furthermore, the dynamic test reveals the need for determining the
following parameters for more detailed model simulations:
- the effective thermal conductivity within the store λ_{eff};
- the flow rate, transfer fluid and storage temperature dependency of
  $(UA)_{f,s}$;
- the distribution of the heat loss over the surface of the store;
- an indication of the occurance of thermosyphoning in the connecting
  pipes.
A calculation model which includes an auxiliary heater can be validated
additionally with respect to:
- the auxiliary heater overall coefficient of heat transfer $(UA)_{f,s,aux}$,
  if a fluid heat exchanger is present;
- the configuration of the model with respect to the number and location of
  segments in the storage material and heat transfer fluid attached to the
  auxiliary heater.

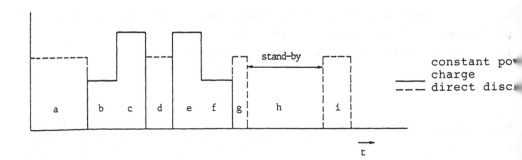

Figure 5.1: Schematic of the two day dynamic test sequence.

A schematic of the dynamic test sequence is shown in Figure 5.1.

It consists of the following phases:

a. stabilization at ambient temperature;

b. half power charge;

c. full power charge;

d. complete discharge to ambient;

e. full power charge;

f. half power charge;

g. half volume discharge;

h. stand-by;

i. complete discharge.

Depending on the kind of information desired, the power charges should be carried out by using either the principal or the auxiliary heat exchanger. If an electric heater is present it may or may not be in operation. During phase (a) the electric heater must always be switched off.

5.4.2. The constant power charges

During the half power phases (b) and (f) the power input should be 5-10 kW/m^3 storage volume. Consequently, during the full power phases (c) and (e) the power input should be 10-20 kW/m^3.

The flow rate during these constant power phases should be the same used in the charge step tests: it should be in line with the producer's recommendations or calculated according to equation (2.1).

The time for all the constant power phases is two-thirds of the charge fill time based on a storage temperature increase of 40 K:

$$t'_{cf} = \frac{40.C_s}{\dot{Q}_c} \qquad (5.1)$$

where $\dot{Q}_c$ is the constant heating power supplied to the store.

The purpose of stabilization phase (a) is to bring the store to a known initial steady state, preferably without heat losses. The half power charge for 2/3 t'_{cf} followed by a full power charge for 2/3 t'_{cf} is equivalent to charging a store with no thermal losses with a perfect heat exchanger.

If there is a potential for stratification within the store it will be built up during phase (b) and enhanced in phase (c).
In the phases (e) and (f) the store is charged to the same level but in the opposite sense. The thermal stratification within the store should be less than before because the reduced power level in phase (f) will give rise to a lower storage inlet temperature. Hence de-stratification may take place.

5.4.3. The complete discharge

The flow rate during the complete (direct) discharge phase (d) should be 0.08 - 0.25 kg/s. The final steady state temperature should be around ambient. Mains cold water may also be used in the direct discharge step. The magnitude of the outlet temperature and the amount of energy withdrawn from the store are used to confirm the correctness of $(UA)_{f,s}$. The discharge temperature profile is used to confirm the modelling of the thermal stratification within the store: the model segmentation and $\lambda_{eff,dd}$.

5.4.4. The half-volume discharge - stand-by - complete discharge

During both the partial (direct) discharge phase (g) and during the complete discharge phase (i) after the stand-by period, the flow rate should be 0.08 - 0.25 kg/s. The inlet temperature during these phases should be around ambient. Mains cold water may also be used in these direct discharge steps. The stand-by period should be 12-24 hrs.
The half-volume discharge often produces a thermally stratified store with hot fluid above cold. During the stand-by period various loss mechanisms take place such as heat loss due to thermal bridges and thermosyphoning. Moreover, the degree of thermal stratification is reduced by heat exchange between the hot and the cold layer. The complete discharge temperature profile enables the model to be judged with respect to the heat loss mechanism, the distribution of $(UA)_{s,a}$ over the storage surface, and the effective thermal conductivity during stand-by, $\lambda_{eff,sb}$, as well as under flow conditions, $\lambda_{eff,dd}$.

5.5. Measurements

The measurements listed in Section 2.5. should be taken. If the dynamic
test is carried out with an electrical auxiliary heater in operation, the
power supplied to the store by this device should also be measured.
A logging rate of 1 minute instantaneous is recommended for the entire
dynamic test.

5.6. Data analysis

5.6.1. Model validation

The calculation model is judged on the basis of the difference between the
experimental and calculated values of the charged or discharged heat and of
the outlet temperature of the store. The data analysis is performed
separately for the charge and discharge periods. If the energy balances and
the temperature deviations meet the criteria, the calculation model is
considered to be valid.

The calculation model always has to contain a transfer fluid and storage
temperature dependent equation for the $(UA)_{f,s}$-value of a heat exchanger,
like equation (2.29). This is because for any situation differing from a
charge temperature step, the $(UA)_{f,s}$-value cannot be considered
constant. With a variable flow, the $(UA)_{f,s}$-value has also to be
considered a function of flow rate. To calculate the $(UA)_{f,s}$-value, the
model requires the input of constants like $c_0 - c_3$ in equation (2.29).
If a model with a certain $(UA)_{f,s}$-formula passes the criteria for
validation set for the dynamic test, then the $(UA)_{f,s}$-formula is
considered acceptable for the range of conditions used in the tests. If the
model fails, then alternative formulae should be tried, based either on
external data or on information derived from a series of temperature step
tests performed at different flow rates, temperature step sizes and
temperature levels.

5.6.2. <u>Energy balances</u>

The relative error between the calculated and measured energy supplied to
or withdrawn from the store is given by:

$$E_E = \left| \frac{Q_{s,c} - Q_{s,m}}{Q_{s,m}} \right| \qquad (5.2)$$

where $Q_{s,c}$ is the calculated heat content. The relative energy error is
determined separately for all charge phases (b)-(c) and (e)-(f) and
discharge phases (d), (g) and (i).
For validation of the model all energy errors E_E must be less than 5%.

5.6.3. <u>Temperature deviations</u>

In addition to the above energy balances criterion, the model is also
judged by the deviation between the calculated and measured storage outlet
temperatures. The standard deviation between the calculated storage outlet
temperature, $T_{e,c}$, and the measured one, T_e, over a certain time
interval (i = 1 $\rightarrow$ N) is calculated from:

$$\sigma(T_e) = \left(\sum_{i=1}^{N} \frac{(T_{e,c,i} - T_{e,i})^2}{N} \right)^{\frac{1}{2}} \qquad (5.3)$$

The criterion for model validation is based on the calculated change in
storage temperature from $\bar{T}_{s,c}(t_1)$ to $\bar{T}_{s,c}(t_2)$ during a charge or
discharge period. This relative storage outlet temperature error is defined
by:

$$E_T = \frac{\sigma(T_e)}{\left| T_{s,c}(t_2) - T_{s,c}(t_1) \right|} \qquad (5.4)$$

Again, the relative temperature error is determined separately for all
charge and discharge phases. The calculation model is regarded as validated
if all relative temperature errors are less than 5%.

6. EVALUATION TECHNIQUES

6.1. Introduction

The description of the tests in the preceding Chapters indicates that it is
impossible to process the measurements manually. That is why various
computer evaluation techniques have been developed.

A distinction must be made between evaluation techniques which analytically
process the experimental data to obtain values for the thermal properties
of the store and so-called identification techniques which compare the
measured data with values resulting from calculations made with a model. In
the latter techniques, the heat storage quantities to be determined are
model parameters. These parameters are identified by a process whereby they
are adjusted iteratively to minimize the squared difference between the
storage outlet temperature obtained form the simulation model and the
measured one. The simulation model is also available in a version which
allows validation to be made by comparison of measured and calculated
results from the dynamic test.

The possibilities of these evaluation techniques are described briefly in
the following Sections. More extensive descriptions can be found in the
user guides in the Appendices. All the evaluation programs are available
on diskettes for IBM AT^R -compatible computers.

6.2. The STEP program

The Storage Test Evaluation Program STEP enables the following quantities
to be determined analytically from a temperature step test:
- the overall coefficient of heat transfer between the store and the
  ambient environment under flow conditions $(UA)_{s,a}(+ (UA)_{f,a})$;
- the heat content and heat storage capacity, $Q_{s,m}$ and $C_{s,m}$
  respectively;
- the heat storage efficiency $\eta_{s,c}$ or $\eta_{s,d}$.

The overall coefficient of heat transfer between the store and the ambient
environment during a stand-by period, $(UA)_{s,a,sb}(+ (UA)_{f,a,sb})$ can be
derived with STEP from a stand-by heat loss test.

Finally, STEP can be used to determine the overall coefficient of heat
transfer between the upper part of the store and the ambient environment,
$(UA)_{s,a,top}$, from a stand-by heat loss test after partial discharge.
The calculation of these quantities from the experimental data is described
in Chapters 2-4. Information about the required format for the measured
data and the control of STEP is given in the user guide in Appendix C.

6.3. The IDENUA program

The IDENUA program is an identification program suitable for determining
the overall coefficient of heat transfer between heat transfer fluid and
storage material from a temperature step test.
The $(UA)_{f,s}$- or $(UA)_{f,s,aux}$-value is found by means of an automatic
search method similar to a bi-section method. The procedure compares
measured values of the heat exchanger outlet temperatures with those
calculated with the simulation routine 4PORT built into the IDENUA
program. A description of 4PORT is given in Appendix D.
In 4PORT each individual fluid segment of the heat exchanger is considered
to be thermally mixed. The conventional definition of $(UA)_{f,s}$, however,
is based on the assumption that heat exchangers have a thermally stratified
(plug flow) response - which they often have. Therefore, the $(UA)_{f,s}$-value
resulting from IDENUA has to be corrected according to equation (2.30).
The IDENUA program is self-explanatory.

6.4. The MRQ-4PORT program (provisional evaluation technique)

Although MRQ-4PORT is a powerful identification program, it has not been protected against
erroneous use and is, therefore, presented here with a certain reserve.
MRQ-4PORT can be used to determine all heat storage quantities. However, it is recommended
that it is used only for those quantities which cannot be determined by STEP - such as the
overall coefficients of heat transfer between heat transfer fluid and storage material,
$(UA)_{f,s}$, and/or $(UA)_{f,s,aux}$, and between storage segments, $(UA)_{s,s,ff}$ and/or
$(UA)_{s,s,dd+sb}$. MRQ-4PORT has also been found to be useful for effective heat exchanger
volume or capacity determination along with, for example, the $(UA)_{f,s}$-value
identification: heat exchanger capacity cannot always be estimated from manufacturer's data.

The Marquardt identification algorithm is used to find the correct parameter values. A set of parameters can be determined at one time. Again, the routine 4PORT is the store model of which the parameters are identified. Storage outlet temperatures calculated by 4PORT are compared with the measured ones.

For reasons explained in Section 6.3, when 4PORT is used for determining the $(UA)_{f,s}$- or $(UA)_{f,s,aux}$-values, these values need to be corrected according to equation (2.30). The user guide of MRQ-4PORT is included in appendix D.

6.5. The DYNAMI program

The DYNAMI program is for processing the measured data from the dynamic test described in Chapter 5. The output of DYNAMI qualifies the simulation model of store under test. As with the programs discussed in the previous Sections, the 4PORT routine has been built into DYNAMI. The calculated storage outlet temperatures are compared with the measured ones for model validation purposes. The energy balances E_E (equation (5.2)) and temperature deviations E_T (equation (5.4)) are determined for all charge and discharge phases.

As explained above, the 4PORT routine considers each heat exchanger segment to be thermally mixed. Hence DYNAMI derives a $(UA)_{f,s}^{model}$-value according to:

$$(UA)_{f,s}^{model} = N_f \dot{m} c_p \left(e^{c_0 \cdot (\dot{m} c_p)^{c_1 - 1} \cdot (T_i - T_s)^{c_2} \cdot ((T_i + T_s)/2)^{c_3}/N_f} - 1 \right)$$

$$(6.1)$$

in which the constants $c_0 - c_3$ are input parameters. See also Section 2.6.5.

The DYNAMI program is self-explanatory.

7. SET OF TESTS FOR HEAT STORAGE CHARACTERIZATION

7.1. Introduction

The decision on which tests are to be carried out for heat storage
characterization depends on the kind of information desired. A distinction
can be made between a simple test method suitable for product
information/selection and an extensive test program which yields parameters
for physical relationships and is suitable for product optimization and
model simulation. However, the boundary between these basic and detailed
characterizations is not always clear.
Section 7.2. gives the minimum set of tests which have to be carried out to
achieve a basic characterization of the store under consideration. Section
7.3. gives guidelines for the derivation of additional quantities from
these tests and for carrying out further tests. Finally, Section 7.4.
outlines the role of the dynamic test.

7.2. Tests for basic characterization

For the basic characterization of a sensible heat storage device, the
following combination of tests should be carried out at least:
- a charge temperature step test from ambient temperature to 50°C, and
- a discharge temperature step test from 50°C back to ambient temperature.
In case of a latent heat store, a series of charge and discharge
temperature step tests should be performed.
All the test conditions for both types of heat store are described in
Section 2.4. The storage device should be operated in the normal charge and
discharge mode according to the manufacturer's instructions.
In the first instance, the following quantities should be determined from
the charge and discharge temperature step tests:
- $(UA)_{s,a,ff}$ (in fact $(UA)_{s,a,ff} + (UA)_{f,a,ff}$);
- $C_{s,m}$;
- $\eta_{s,c}$ or $\eta_{s,d}$;
- $(UA)_{f,s}$ (if a heat exchanger is present).

The heat storage capacity determined from the charge and discharge tests should not differ more than 5%. If the difference is larger, the cause has to be tracked down. For instance, the stabilization times may be too short, there may be dead zones in the store where heat transfer only takes place by way of thermal conduction and there may (in a latent heat store) be supercooling.

7.3. Tests for detailed characterization

7.3.1. Criterion for determination of the stand-by heat loss

If the heat loss during stand-by is expected to be substantially different from the heat loss at finite flow rate determined from a temperature step test, the stand-by heat loss test described in Chapter 3 should be carried out. Lower stand-by heat losses may, for instance, occur in a mantle type heat exchanger when a heat transfer fluid like air acts as an insulator during stand-by periods.

7.3.2. Criterion for determination of the heat loss distribution over the store surface

If the measured $(UA)_{s,a}$-value determined from a temperature step test is substantially (e.g. more than 2 times) larger than expected on the basis of the geometry of the store and the type and thickness of the insulation material and if thermal stratification of the store is possible, then the partial discharge - stand-by heat loss test should be carried out to determine $(UA)_{s,a,top}$.

7.3.3. Criterion for determination of the effective thermal conductivity within the store under finite flow conditions

From either a charge/discharge temperature step test using a high performance heat exchanger distributed over the whole store or a direct charge/discharge of the storage volume, the $\lambda_{eff,ff}$-value (for the storage model converted into $(UA)_{s,s,ff}$) can be determined (see Section 2.6.6).

The procedure is as follows:

- run the step test with an appropriate identification method which uses a mathematical model of the store. Start with $(UA)_{s,s}^{model}$ = 0 W/K as fixed input value; the segmentation of the store model should obey equation (2.25);

- compare the calculated step response curve with the measured outlet temperature curve. Then,

  . if the absolute value of the temperature gradient dT_e/dt of the measured curve is larger than that of the simulated curve as shown in Figure 7.1, then the storage behaviour can be considered to be fully stratified and there is no need for a more precise identification. In which case, the following holds:

$$(UA)_{s,s,ff} \leq (UA)_{s,s,ff}^{num} \qquad (7.1)$$

or, in other terms:

$$\lambda_{eff,ff} \leq 0.5 \ (1-f) \ \dot{m}c_p \ \frac{L_s^2}{N_s V_s} \qquad (7.2)$$

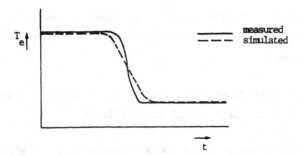

Figure 7.1: Example of a step response where the absolute value of the measured outlet temperature gradient is larger than the simulated one.

. if, on the other hand, the absolute value of the temperature gradient
dT_e/dt of the measured curve is smaller than that of the simulated
curve as shown in Figure 7.2, then $(UA)_{s,s,ff}^{model}$ should be identified to
get a good fit. After that, the $\lambda_{eff,ff}$-value is calculated according
to the equations (2.35) and (2.34).

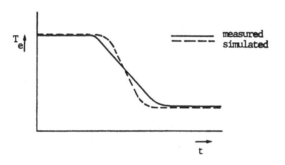

<u>Figure 7.2</u>: **Example** of a step response where the absolute value of the
measured outlet temperature gradient is smaller than the
simulated one.

7.3.4. <u>Criterion for determination of the effective thermal</u>
<u>conductivity within the store during stand-by</u>

From a partial discharge - stand-by heat loss test followed by a complete
discharge as described in Chapter 4, the $\lambda_{eff,sb}$-value can be
determined. The $\lambda_{eff,dd}$-value has to be known already. The procedure is
similar to that described in Section 7.3.3:
- run the test with an appropriate identification method which uses the
  mathematical model of the store. Start with $(UA)_{s,s}^{model} = 0$ W/K as fixed
  input value; the segmentation the store model should obey equation
  (2.25);

- compare the calculated step response curve with the measured outlet
  temperature curve. Then,
  . if the absolute value of the temperature gradient dT_e/dt of the
  measured curve is larger than that of the simulated curve as shown in
  Figure 7.3, then the store can be considered to be fully stratified and
  there is no need for a more precise identification.
  In which case, the following holds:

$$(UA)_{ss,dd+sb} \leq \frac{t_d}{t_d + t_{sb}} (UA)_{s,s}^{num} \tag{7.3}$$

or, in other terms:

$$\lambda_{eff,dd+sb} \leq 0.5 \ (1-f) \ \dot{m}c_p \ \frac{L_s^2}{N_s V_s} \ \frac{t_d}{t_{sb} + t_d} \tag{7.4}$$

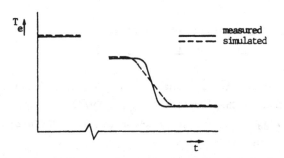

Figure 7.3: Example of a temperature response of a partial discharge –
             stand-by heat loss test where the absolute value of the
             measured outlet temperature gradient is larger than the
             simulated one.

. if, on the other hand, the absolute value of the temperature gradient dT_e/dt of the measured curve is smaller than that of the simulated curve as shown in Figure 7.4, then $(UA)_{s,s,dd+sb}^{model}$ should be identified to get a good fit. After that the $\lambda_{eff,dd+sb}$-value is calculated according to the equations (4.7) and (4.6) and the $\lambda_{eff,sb}$-value according to equation (4.8).

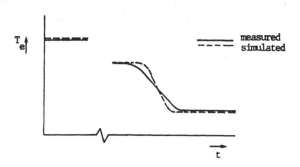

Figure 7.4: Example of a temperature response of a partial discharge - stand-by heat loss test where the absolute value of the measured outlet temperature gradient is smaller than the simulated one.

7.3.5. Criteria for auxiliary heater tests

For the characterization of a heat exchanger for auxiliary heating, an auxiliary heater temperature step test as described in Chapter 2 should be carried out.

The correctness of the modelling (i.e. the positioning) of a built-in electrical auxiliary heater should be assessed using a dynamic test with the electric heater in operation (see Section 7.4).

7.4. The dynamic test as final test

After the tests for basic and (if necessary) detailed characterization of
the store have been carried out according to Sections 7.2 and 7.3, the
dynamic test has to be carried out as final check. This reveals the
correctness of the measured quantities as model parameters as well as the
need for any additional tests. If the dynamic test results are
satisfactory, the calculation model of the heat storage device can be
regarded as sufficiently accurate. If they are not satisfactory, the model
or model parameters have to be changed.

References

[1] Visser, H. and H.A.L. van Dijk (ed.)
    Final Report on the Activities of the Solar Storage Testing Group.
    Commission of the European Communities, EUR 13119.

[2] Kübler, R., M. Bierer and E. Hahne
    Heat transfer from finned and smooth tube heat exchanger coils in hot
    water stores.
    Advances in solar energy technology, volume 2, page 1177-1181.
    Proceedings of the biennial congress of the International Solar Energy
    Society, Hamburg, FRG, September 13-18, 1987.

APPENDIX A:

RECOMMENDATIONS FOR TEST FACILITY DESIGN AND FOR MEETING THE REQUIRED
MEASURING ACCURACY

CONTENTS

1. <u>INTRODUCTION</u>

Appendix A gives recommendations for test facility design and for meeting the required accuracies for data measurement. The discussion is confined to issues relating specifically to the testing of heat stores. For general matters on, say, test facility control, measurement accuracy or calibration of sensors, the researcher is referred to handbooks dealing with these subjects.

An example of a test facility is given in Chapter 2. In this chapter, also, will be found a discussion of the quality of the store inlet temperature step. Chapter 3 pays special attention to the improvement of the accuracy of measurement of the temperature difference across the store as well as to the checking of the validity of calibration results by use of a reference heater.

2. RECOMMENDATIONS FOR TEST FACILITY DESIGN

2.1. General

The test environment should have a constant temperature and therefore at
the very least should be protected from direct insolation. Figure 2.1 shows
an example of a test facility.

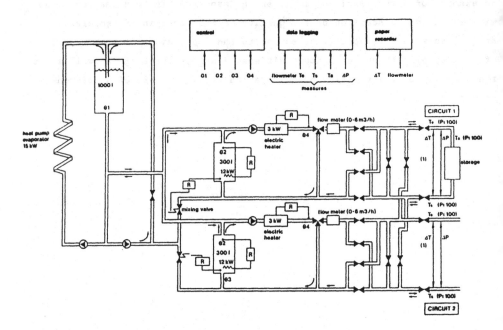

Figure 2.1: Example of a test facility.

The piping between the outlets of the heat transfer fluid temperature
regulators and the inlets to the storage device should be as short as
possible to minimize the effects of the surrounding environment on the
fluid inlet temperature. The connecting hoses or pipes must lead downwards
from the store under test to the test facility. There should be a valve
adjacent to the store in each loop of the system. To prevent
thermosyphoning, the valves in those loops which are not being used (and,
hence, through which no fluid is flowing) must be kept closed.

All piping should be corrosion-resistant and suitable for use at temperatures up to 100°C. If special non-aqueous fluids are used then care should be taken to ensure that the pipes are compatible with them.

The pump in the fluid loop should be located in such a position that the heat which it dissipates into the fluid does not impair either the control or the measurement of the storage inlet temperature.

Filters (nomial filter size 200 μm) should, at the very least, be placed upstream of the flow rate measuring device and the pump.

The heat transfer fluid used for the test may be water or a fluid recommended by the store manufacturer. The specific heat and density of the fluid should be known to an accuracy greater than 1% over the range of temperatures used in the tests.

Some fluids may need to be changed periodically to ensure that their properties remain well defined. A drawback to using a fluid other than water is that its properties are often not known to the same degree of accuracy. If there is a choice, therefore, water is to be preferred. When the test loop is freshly charged with water the system should be heated to its maximum operating temperature to expel air dissolved in the water.

2.2. Quality of the inlet temperature step

The most well-defined temperature step change is achieved when two separate loops are available. One is used to maintain the initial temperature of the store. The other is adjusted to the final temperature using a bypass to the store. When the moment for the step is reached, the first loop is closed and the second one opened.

When only one test loop is available, its thermal capacity must be small so that a rapid temperature change can be achieved. Just before the temperature step the storage device may be bypassed to change the temperature in the test loop. This temperature change must be rapid (within a few minutes) to maintain a well-defined initial temperature for the storage device, which will cool off during this period.

The time required to reach 95% of the temperature change is shorter for a double loop system or a single loop system with a bypass than for a single loop system without a bypass. Moreover, the average inlet temperature of the store during the change is closer to its final value.

3. RECOMMENDATIONS FOR MEETING THE REQUIRED MEASURING ACCURACY

3.1. General remarks about the calibration procedure

The actual accuracy of measurement depends on a chain which starts with the
physical parameter and proceeds via the sensor, transducer and data logger
through to the data processing. The quality of this chain can be
established by either:

a. testing each component separately, or

b. testing the complete chain as a whole.

The advantages of the first approach are:

- well-known relationships (e.g. between temperature and thermocouple
  output or between the input and output points of a data logger) can be
  brought into play to interpolate between points at which calibration is
  carried out. Thus, the number of calibration points in a particular
  temperature, voltage, etc., range can be kept to a minimum.
- the reliability of a single component (e.g. data logger) can be checked
  without the need for recalibration of the complete chain.

The obvious disadvantage of the first approach is that a calibration
procedure has to be carried out for each component.

The advantages of the second approach are:

- the chain from physical parameter to output reading can be considered as
  a 'black box';
- the number of instruments to be calibrated is restricted.

The disadvantage is that a complete scan is required over the whole range
of conditions (e.g. flow, temperature) in order to establish the
relationship between input and output data.

If the researcher suspects that there might be some interaction or
disturbance between successive parts of the chain, then he should opt for
the second approach.

3.2. Measurement of temperature difference across the store

3.2.1. Sensor heat losses and temperature profiles

Because a high accuracy is required for measurement of the temperature
difference across the store, special attention must be paid to avoiding
systematic errors due to sensor heat losses and their temperature profiles.
The sensors are mounted on the ports of the store but are meant to measure
temperature in the ports. The housings of the sensors have certain heat
losses which give rise to sensor temperature readings which are higher or
lower than the actual temperatures in the inlet and outlet ports. To limit
these heat losses the two parts of the differential temperature sensor
should be located as close as possible to the inlet and outlet of the
store. Care over positioning these probes is even more important than for
the absolute temperature probes (see Figure 3.1). Heat leakage along
connecting leads can be minimized by using thin leads, insulation and
adequate depths of immersion.

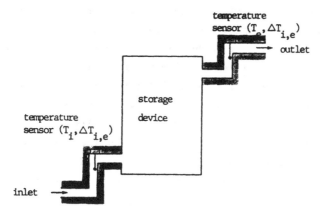

Figure 3.1: Recommended sensor positions for measuring the heat transfer
fluid temperature difference across the store.

The heat losses of the housing and upstream piping of the sensors cause
temperature profiles in the sensors' cross-sections. The mean cup
temperature is therefore lower than the center temperature sensed by the
probes.
In addition, self heating effects of PT 100 transducers and stem effects
(i.e. heat losses through connecting wires and stem) may influence the
measured temperatures and lead to errors.

When measuring small temperature differences the errors of the two parts of
the ΔT-sensor may cancel each other out.

To minimize the possible error, the fluid should be mixed effectively at
the position of the sensors. To achieve this, a bend in the pipework or an
orifice should be placed upstream of each sensor and the sensor probes
should point upstream (see Figure 3.1).

3.2.2. Offset tests on the sensors

Even if all possible precautions are taken to avoid errors due to heat
losses and temperature profiles in the sensors, there may remain some
systematic errors in the sensor measurements. These systematic errors will
not be detected by direct calibration against reference instruments because
both the reference instrument and the instrument being calibrated will be
based on the same knowledge and technology. These matters will always be a
topic for discussion. However, the possibility of systematic errors can be
overcome by carrying out offset tests.

Offset tests should be carried out in the test configuration by replacing
the storage device by a short piece of well-insulated piping and measuring
the temperature difference as a function of temperature and flow. These
offset tests should be carried out over the range of fluid flow rates and
temperatures to be used during testing of the store itself. They act as a
combined check on the zero reading of the temperature difference sensor and
on the effect of heat losses and temperature profiles.

Measurement of the heat loss of stores is the test which puts the greatest
demand on the accuracy of the sensors. This is because in this test both
parts of the ΔT-sensor are typically on the same temperature level (with
respect to the ambient temperature).

In the offset tests the temperature difference sensor should read ΔT = 0 so
that any offset from ΔT = 0 at a specific temperature and flow can be used
to correct the heat loss ΔT measurement carried out under the same
temperature/flow. This means that the upstream piping and its insulation as
well as the insulation of the temperature difference sensor itself should
be identical to those in the storage test loop. Also the temperature
profile in both parts of the temperature difference sensor must be the same
for the offset and the storage tests.

The recommended tests and their sequence are given in Table 1.

Table 3.1: Offset tests on ΔT-meters placed in series in one circuit.

test nr	flow rate [1/hr]	temperature [°C]
1	50	70
2	100	70
3	200	70
4	400	70
5	800	70
6	800	45
7	400	45
8	200	45
9	100	45
10	50	45

Test 1 at least, if not other tests, should be repeated with the two parts
of the transducer exchanged. The 'reverse' test should reproduce the error
measured in the corresponding 'forward' test. If this is so, then the error
is systematic and any measurement of the temperature difference at a
certain temperature and flow rate can be adjusted with the systematic
error found by interpolation of the offset test results.
If the 'reverse' test does not give the same offset as measured earlier,
then there must be differences between the two parts of the ΔT-sensor. For
instance:
- the insulation of the two parts may not be identical;
- the length and/or positioning of the probes may not be identical;
- the flow pattern through the two parts may not be identical.
The irreproducable part of the offset is to be added to the measurement
uncertainty of the transducer.

3.2.3. Improving the accuracy

From the above discussion, it can be seen that the following measures
should be taken to improve the temperature difference (and temperature)
sensors:
- improve the insulation of the sensors and the upstream piping;
- design the dimensions of tubes, etc., to ensure turbulent flow at all
  flow rates and install flow mixers;
- place the probes exactly in the centre of the cross-section of the flow
  channel.

The latter can be achieved by using plastic insert tubes to hold the
probes. This narrows the bore, (and therefore achieves turbulent flow) and
the plastic tube also improves the insulation. Different inserts can be
used for different flow ranges to minimize the pressure drop.

3.3. Cross checks with a reference heater

3.3.1. General

A reference heater can be used to check the validity of the calibration
results. Reference heater tests allow an independent judgement to be
obtained on the accuracy of the total monitoring system including
ΔT-measuring equipment, flow meter and data acquisition and processing
system. Possible errors and their causes can be identified. Reference
heater tests also provide a check on a need for recalibration: if they give
good results, recalibration of the measuring equipment in the test loop is
not required.

An inventory of reference heaters used by several institutes was carried
out within the framework of IEA Solar Heating and Cooling Programme Task
III 'Performance testing of solar collectors' [1]. The participants in the
SSTG used a reference heater designed by the Thermal Insulation Laboratory,
Denmark [2]. A schematic illustration of this is given in Figure 3.2.

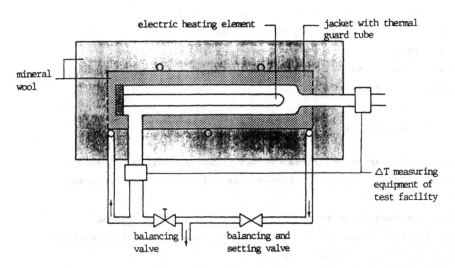

Figure 3.2: Schematic illustration of the reference heater used in the
SSTG.

The reference heater is a well-insulated electric heating element which can replace a storage system in the test loop. The difference in heat content between inlet and outlet measured under normal operating conditions is compared with the measured electric power consumption of the reference heater. To ensure an accurate check, the heat loss over the reference heater should be very low. If correction of the heat loss is possible, this is even better. A precision Watt meter should be used to measure the power supply.

The cross check of the accuracy of the test loop with the reference heater must be carried out using exactly the same insulation, sensors, data logger and data processing routine as is used for normal storage tests.

The reference heater may be used at regular intervals between test series.

3.3.2. Reference heater tests

In performing the reference heater tests with a heater similar to the one used in the SSTG, the operating instructions given in [2] should be followed. A set of basic tests is specified in Table 3.2. If the basic tests throw up significant discrepancies, additional tests should be carried out under different conditions to identify the cause.

Table 3.2: Basic reference heater tests.

test nr	inlet temperature °C	flow rate l/hr	power supply W
1	ambient	50	1500
2	ambient	800	0
3	ambient	800	1500
4	70	800	1500
5	70	800	0
6	70	50	0
7	60	50	1500

3.3.3. Analysis of the reference heater test results

One or more tests may reveal a discrepancy between the measured power
supply from the reference heater and the power consumption derived from
flow and ΔT measurement. Depending on the test at which the discrepancy
appears, its cause can be identified to a greater or lesser extent.

Test 2:

In this test, heat loss will have a negligible effect on $\Delta T = 0$. Therefore,
any deviation from $\Delta T = 0$ will be caused by an offset in the ΔT-reading.

Test 5:

Here, heat loss will induce a small ΔT. When this is taken into account
(see Test 6 for derivation of the heat loss rate) any remaining ΔT can be
said to be caused by an offset in the ΔT-reading.

Test 6:

The overall coefficient of heat transfer between reference heater and
ambient environment can be derived from this test using the following
equation:

$$(UA)_{rh,a} = \frac{\dot{m}c_p \ (T_i - T_e)}{0.5 \ (T_i + T_e) - T_a} \tag{1}$$

where T_i and T_e are the inlet and outlet temperatures of the reference
heater, $\dot{m}$ and c_p the mass flow rate and specific heat capacity of the
fluid and T_a is the ambient temperature.

Tests 1 and 3:

These tests provide a caloric check with negligible influence from heat
loss. In Test 3, the temperature difference will be small, which gives an
inevitable uncertainty in the result - up to 4 percent, if the required
accuracies in ΔT (0.05 K) and flow rate (1%) are met.

Tests 4 and 7:

These tests also provide a caloric check, but here the heat loss has to be
taken into account.

REFERENCES

[1] IEA, SHEC, Task III, Performance testing of solar collectors, Reference and Calibration Heaters, Swedish Council for Building Research, ISBN 91-540-4501-0, January 1986.

[2] Furbo, S. and S. Svendsen, Description and Operation of the Reference Heater, TIL, Technical University Denmark, January 1987.

APPENDIX B:

ACCURACY AND CONFIDENCE IN TESTING AND TEST EVALUATIONS

CONTENTS

1. <u>INTRODUCTION</u>

Appendix B deals with error analysis and related topics relevant to an
understanding of the credibility of test results.
A brief introduction to the main concepts of error analysis (Chapter 2) is
followed by an explanation of error estimation in the evaluation of a
temperature step test. This includes an explanation of the uncertainty
bands given in the output from STEP.
Chapter 4 summarizes the main results of a study aimed at finding the
sensitivity of the annual performance of a solar energy system to the
uncertainty in each of the thermal characteristics of the store. This
indicates the importance of an uncertainty in any one of the
characteristics derived from the tests.
Finally, Chapter 5 contains some points worthy of attention on the subject
of model validation.

2. ERROR ANALYSIS

2.1. Definitions

2.1.1. General

The measured value of a quantity is a more or less precise and accurate
estimate of its true value. The measurement is precise if the difference in
individual results from repeated measurements is small. The measurement is
accurate if the mean value of repeated measurements is close to the true
value. A precise measurement is achieved if the random errors are small;
an accurate measurement requires the systematic errors to be small.

2.1.2. Random errors

A random error is a difference between a measured and true value which is
independent of the measurement conditions. The frequency distribution of
this deviation approaches a certain shape after a high number of
independent measurements, no matter who is making the measurements and
where and when they are done. In other words, the measurements are
independent. Often the frequency distribution will resemble the gaussian
curve (normal distribution), but this is not necessarily so.

2.1.3. Systematic errors

A systematic error is a deviation from the true value which is duplicated
exactly at each repetition of the measurement under the same measurement
conditions. In general the size of the systematic error depends on the
test conditions (who, when and where).

2.2. Confidence interval

2.2.1. General

In error analysis we deal with errors of more or less unknown size.

It is this which brings us to the definition of confidence intervals: the
probability that the true value lies within the P percent confidence
interval around the measured value is P percent. The 95 percent confidence
interval is often the most convenient one. If the interval is symmetrical
around the measured value one can also use the definition that the
probability that the measured value deviates less than half of the interval
from the true value is P percent.

2.2.2. Random errors

For a <u>random error</u>, the standard deviation σ is the measure which indicates
the width of the distribution curve.

For a <u>normal</u> distribution the probability that the measured value deviates
less than 2σ (to be precise: 1.96σ) is 95 percent. So the 95 percent
confidence interval is 2σ.

For other probability distributions this statement is not valid - at least
not for an <u>individual</u> measurement.

A special characteristic of the random error with <u>normal distribution</u> is
that the probability distribution of the mean value of independent
measurements is narrowed with a factor $\sqrt{n}$.

However, the probability distribution of the <u>mean</u> value of the error from a
<u>high number</u> (minimum - 30) of <u>independent</u> measurements becomes gaussian,
regardless of the probability distribution of the individual errors. This
implies that for <u>any</u> probability distribution a confidence interval can be
estimated when a sufficient number of individual measurements have been
taken. From these measurements the standard deviation σ can be derived and
2σ is the 95 percent confidence interval. There is one condition to this:
the probability distribution of the independent measurements must be the
same - as they would be in the case of repeated measurements on the same
quantity.

2.2.3. Systematic errors

In general, the only parameter which can be used to estimate a confidence
interval is the maximum possible error, ε_{max}: a measured value will
deviate certainly (or with a high level of certainty) less than $\pm \varepsilon_{max}$
from the true value.

This means one can compare the almost 100% confidence interval of 2σ for a random error with the 100% confidence interval ϵ_{max} for a systematic error.
Notice that there is no such similarity for lower levels of the confidence interval.

2.3. Summation of errors

2.3.1. General

Some elements of the summation of errors have been introduced implicitly already - for instance, we have said that the error in the mean value of a number of independent measurements is the result of a summation of individual errors.

2.3.2. Random errors

Due to the statistical character of the random error with its normal distribution it is possible to quantify the probability of a combination of actual values coming from M independent sources.
It can be derived that one should sum the squares of the individual P percent confidence intervals to get the square of the overall P percent confidence interval.

2.3.3. Systematic errors

For a systematic error one should always be prepared for the maximum possible errors occurring: the probability that the individual actual error is close to the maximum value is in general not smaller than the probability of having a smaller value.
However, if we have a high number (> 30) of independent sources of systematic error then we have the same effect as with a random error with non-gaussian distribution (Section 2.2.2). Regardless of the (here unknown) probability distribution, the sum of M independent systematic errors becomes gaussian if M is sufficiently large (M > 30) and if the probability distribution of each of the errors is the same. So in that case the individual maximum possible errors must be summed quadratically to get the overall 95 percent confidence interval.

If the two conditions (i.e. large M and same probability distribution) are
not met, then for the sum of M systematic errors one can only rely on the
'100 percent confidence interval' for each error - i.e. the maximum
possible error. In this case the absolute values of the maximum possible
errors have to be summed linearly; this gives a larger confidence interval,
which reflects the probability that the different errors are not
compensating for each other. In practice, this implies that the linear
summation of only a few independent systematic errors with unknown
probability distribution would give a rather pessimistic confidence
interval. As long as the errors are really independent one could defend a
quadratic summation instead, although there is the (small?) probability
that the resulting interval is too optimistic (i.e. small) for 95 percent
confidence. The main condition, however, is that the errors are
independent.
If these arguments are valid for systematic errors, they are obviously also
valid for a combination of systematic and random errors.

2.3.4. Correlated errors

If errors are correlated, then one should sum (the absolute values of) the
maximum possible or 2σ - values linearly.
A typical situation in which errors may be correlated is where we have a
specific quantity measured as a function of time.
If systematic errors are involved it is obvious that one is not allowed to
assume that the value of the error at certain moment is independent of the
value at the previous observation: the confidence interval of the mean
value of n observations will certainly not be reduced with $1/\sqrt{n}$.
For a random error this reduction of the confidence interval is only
allowed if the successive values are really independent. This is not the
case if the sampling rate is higher than the time constant (inertia) of the
sensor: in this situation the measurements are replicated rather than
repeated.

2.4. References

References [1]-[4] give the most relevant international standards on
statistics with respect to experimental errors.

3. ERROR ESTIMATION IN THE TEMPERATURE STEP TEST RESULTS

3.1. Introduction

The storage test evaluation program STEP contains an estimation of the
possible error in the temperature step results due to uncertainties in the
measured quantities.
This Chapter presents the rationale for the chosen approach and explains
how the presented uncertainty bands should be interpreted.

3.2. Input quantities

In the evaluation of a step test the possible error has to be given for
each of the input quantities. If possible, a distinction should be made
between the random part (expressed in terms of twice the standard
deviation, 2σ) and the systematic part (expressed in terms of the maximum
possible error, ε_{max}). Both parts may be summed to give the overall 95
percent confidence interval.
The minimum sampling rate during successive parts of the test is adapted to
the rate at which the measured quantities typically vary (see Part B
Chapters 2-5). In the integration of the net heating power supplied to the
store, therefore, the influence of the random part of the error in the
measured quantities of ambient temperature, inlet temperature, temperature
difference and flow rate will be strongly reduced. And unless the random
part of the error in these quantities is very dominant its effect can be
neglected.
Our main effort therefore, should be to minimize the maximum possible
systematic error. This error can be made up of a number of elements.
Those which perhaps will require most attention will be a (constant) offset
or a deviation from a calibration curve of too low an order are worthy of
more attention.

3.3. Summation of errors

The number of independent quantities involved in the calculation of
quantities such as the net heating power, storage capacity, storage
efficiency and heat loss rate is not large.

For the summation of the contributions from each of the individual
independent sources, the most pessimistic assumption is made. It is assumed
that there is a high probability of the errors from each individual source
having the maximum possible value and mutually reinforcing each other. The
errors are, therefore, summed linearly.

3.4. Time integration

The time integration of a measured quantity like the net heating power
supplied to the store is a summation of successive values for the same
quantity. Obviously these successive values are not independent so again,
the errors are summed linearly.

3.5. Influence of fluctuations

In some of the equations used in the evaluation of a temperature step test,
mean values of a time series of measurements are involved. This occurs, for
instance, with the ambient temperature in the calculation of the heat loss
capacity rate. Fluctuations around the mean value which add to the
uncertainty in the derived result are taken into account by adding
additional uncertainty to the measured mean value (see Appendix C). The
effect of drift is weighted more strongly than the effect of a random
variation and conforms to the suggestions in Section 3.2 above.

3.6. Heat loss correction

The net heating power supplied to the store is corrected for heat loss from
the store to the ambient environment. If the heat loss capacity rate has
been determined from the steady state condition after or before a step
response test, it can be assumed that the errors in the heat loss capacity
rate and the heat input to the store under steady state are about the same
in size and sign.

Hence the errors in the heat input to the store and in the heat loss
calculated during the step are to some extent counterbalanced in the net
heating power, the difference between the two quantities. This partial
compensation is duly taken into account by neglecting the error in the heat
loss if the heat loss capacity rate has been determined during the steady
state situation before or after the step (see Appendix C).
Otherwise, the errors are summed quadratically, not linearly. This is an
arbitrary compromise to avoid an exaggerated overestimation of the
uncertainty in the final results.

3.7. Conclusion

A number of assumptions had to be made to achieve a fair estimation of the
uncertainty in the step response tests.
The uncertainties reported in the results of the Storage Test Evaluation
Program STEP are rather pessimistic values since they (with one exception
discussed in Section 3.6 above) do not include the possibility that errors
from different independent sources might compensate for each other.
A sensitivity study revealed that an even more strict application of this
rule could lead to an increase of 30 percent in the estimated uncertainty
of the measured storage capacity for a typical case.
On the other hand, the same study also revealed that the assumption of a
normally distributed random variation in the actual size of each individual
independent error (between zero and the maximum possible size) would lead
to a 30 percent lower value for an estimated uncertainty in the storage
capacity.

4. INFLUENCE OF UNCERTAINTIES IN TEST RESULTS ON ANNUAL SYSTEM PERFORMANCE SIMULATIONS

4.1. Introduction

Sensitivity studies were carried out as part of the SSTG activities with the aim of quantifying the influence of inaccuracies in storage test results on the calculated long term thermal performance of solar energy systems.

The main parameters of a storage device model are:

- the storage capacity;
- the storage heat loss;
- the heat exchanger overall coefficients of heat transfer for charging and discharging.

Second order parameters of a storage device model are:

- the capability of building up thermal stratification during charge or discharge;
- the capability of minimizing numerical thermal diffusion, due to the applied calculation technique.

The simulated system is a domestic hot water (DHW) system. This is because more than 90% of the commercially-available solar storage devices are developed for such systems. Moreover, the system performance is likely to be more sensitive to the storage model parameters in case of a DHW system than in a system for space heating. For, the DHW system has a very explicit load profile and demand temperature.

The dimensions of the DHW system roughly correspond to the average dimensions of systems installed in The Netherlands. The main characteristics are:

- storage volume 0.1 m³;
- overall heat loss coefficient of the store: 2.0 W/K;
- overall coefficient of heat transfer between transfer fluid and store: 150 W/K;
- collector area 3 m²;
- effective τα-product 0.7;
- constant part of heat loss coefficient 3 W/m²K;
- temperature-dependent part of heat loss coefficient 0.013 W/m²K²;

- collector efficiency factor 0.95;
- fluid density x specific heat 4.18 kJ/kgK;
- flow rate:
  . high flow when one heat exchanger segment: 0.01 kg/s/m²;
  . low flow when more heat exchanger segments: 0.003 kg/s/m²;
- backup heater:
  A backup heater is provided to heat the tap water to 55°C if necessary.
  This heater is assumed to operate with ideal characteristics;
- hot water consumption per day:
  100 litres heated from 10°C to 55°C.

A full report on the assumptions, method and results can be found in [5], Part B Chapter 12.

4.2. Main results

The influence of uncertainties in the storage parameters is expressed in the number S, the sensitivity:

$$S = \frac{\Delta SF}{SF_{nom}} \bigg/ \frac{\Delta P}{P_{nom}} \qquad\qquad (4.1)$$

where:

SF	Solar Fraction, i.e. the proportion of the heat demand provided by solar energy	(-)
ΔSF	Solar Fraction difference	(-)
SF_{nom}	nominal Solar Fraction	(-)
P	storage parameter	(unit P)
ΔP	error or deviation in a storage parameter from its nominal value	(unit P)
P_{nom}	nominal storage parameter	(unit P)
S	sensitivity of the solar fraction to a deviation in a storage parameter	(-)

With equation (4.1) we can derive the possible error in an annual system performance due to parameter uncertainties. S times a known parameter uncertainty gives the uncertainty of a system's solar fraction.

A compilation of the main results is given in Table 4.1.

These results are valid for typical types of domestic hot water system (see [5], Part B Chapter 12). The last column in Table 4.1 shows how S may be increased if the parameter itself is up to 30 percent larger or smaller than the nominal value combined with some other major changes in the system design and/or operation.

Table 4.1: Main results with respect to the sensitivity S.

parameter	sensitivity S for typical system design	sensitivity S for sensitive situation [1])
overall coefficient of heat transfer between transfer fluid and storage material $(UA)_{f,s}$	0.025	0.06
storage capacity C_s	0.15	0.25
overall coefficient of heat transfer between storage and ambient $(UA)_{s,a}$	- 0.10	- 0.15

[1]) the highest value found with up to 30 percent variation in parameter value in combination with some other system variations.

Example:

If due to uncertainties the relative error of the $(UA)_{f,s}$-value is 50% and the system is sized according to the assumptions made for this study, then the relative uncertainty of the solar fraction would read:

$$\frac{\Delta Sf}{Sf} = 0.025 * 50\% = 1.25\%$$

Note that the sensitivity can be much higher if the system dimensions are changed.

More results can be found in [5], Part B Chapter 12.

Most storage models use a fixed number of storage segments. Such models may
suffer from numerical thermal diffusion of segments during charge or
discharge. This occurs when the (dis)charged volume is not exactly one (or
a multiple of one) segment volume. This process is described by the
parameter f in equation (2.26) in Part B Chapter 2.

Table 4.2 summarizes the influence of the f-value on the prediction of the
solar fraction for a range of system designs and applications. The results
are valid for the store modelled as 10 segments. The decrease in the
calculated solar fraction is compared with a typical decrease in
performance due to a fully mixed discharge.

Table 4.2: Main results with respect to the f-value.

type of store	number of store segments	f-value	decrease in calculated solar fraction compared to ideally stratified discharge case (%)
ideally stratified discharge	10	1	0
	10	0.6	0 - 3
fully mixed discharge	(not relevant)		7 - 10

Hence the requirements for the minimum number of nodes in Part B Chapter 2
(equation (2.25)). More results can be found in [5], Part B Chapter 12.

5. MODEL VALIDATION: SOME POINTS WORTHY OF ATTENTION

5.1. Errors in test results

In the validation of a model against the results of an experiment, the
latter is considered to be the reference, i.e. 'the true model'.
In the use of the dynamic test for validating the storage model there are,
however, three problems:

1. the uncertainties in variables measured during the dynamic test which
   are used as input data for the model calculation;
2. the uncertainties in the measured 'output' of the test, the amount of
   energy supplied to or extracted from the store and the outlet
   temperature;
3. the uncertainties in model parameters derived from previous tests.

Regarding 1, we can say that the complexity of the storage calculation
model does not allow a straightforward mathematical procedure to be used to
estimate the influence of the different sources of error on the results. If
the errors were purely random, their effect could be determined by imposing
a random variation as a simulated error on each of the quantities involved
and observe the effect. However, the errors may for a large part consist of
systematic parts which are either constant or float (drift) during the
test. Only a Monte Carlo technique, involving sampling from all possible
combinations, could help to quantify the effect.

Concerning 2, the uncertainty in the output of the dynamic test should be
as small as possible since it is used as reference for judging the model's
result.

Concerning 3, the influence of the uncertainties in the model parameters
can be found easily by modifying the values fed into the model and
observing the effect on the result.

Another mechanism which may disturb the validation is cancelling of errors
due to correlation between (systematic) errors during the tests for the
derivation of the storage characteristics and errors during the dynamic
test. For instance, if the flow rate reading is systematically too high,
then the derived value for thermal capacity and heat loss capacity rate
will be too high.

If the dynamic test is carried out in the same test loop with the same
overestimation of the flow rate and the model is fed with overestimated
capacity and heat loss capacity rate, then the measured and calculated
energy transfer and outlet temperatures may well agree.

5.2. The root mean square difference as criterion

The results of the test and the model calculation are compared by taking
the root mean square (RMS) deviation of the outlet temperature for
successive parts of the dynamic test.
A typical problem may occur when there is a step gradient in the output: it
is possible that the model calculates the gradient very accurately, but
with a small shift in time. An RMS deviation may in that case give too
pessimistic an impression of the result.
Another problem may occur if the model suffers from the so-called
'numerical diffusion'. If the guidelines for the minimum number of segments
(Part B equation (2.25)) are met the calculation model will predict the
effect of a thermally stratified response, if present, accurately enough.
Nevertheless, the calculated outlet temperature may in fact be more
'smeared' than the measured one, so that there is a large RMS deviation in
outlet temperature.

REFERENCES

[1] ISO 3534, Statistics-Vocabulary and symbols; 1977-07-01.

[2] ISO 2602, Statistical interpretation of test results - Estimation of
    the mean - Confidence interval; 1980-02-15.

[3] ISO 3207, Statistical interpretation of data-determination of a
    statistical tolerance interval; 1975-05-15.

[4] ISO 2854, Statistical interpretation of data-techniques of estimation
    and tests relating to means and variances; 1976-02-15.

[5] Visser, H. and H.A.L. van Dijk (ed.)
    Final Report on the Activities of the Solar Storage Testing Group.
    Commission of the European Communities, EUR 13119.

APPENDIX C:

STORAGE TEST EVALUATION PROGRAM VERSION 2.2*)

CONTENTS

*) prepared by R. Kübler, Institut für Thermodynamik und Wärmetechnik,
   University of Stuttgart, FRG.

1. <u>INTRODUCTION</u>

The Storage Test Evaluation Program (STEP) was developed to guarantee that
the storage tests are evaluated exactly in line with the equations proposed
by the SSTG. Inevitably, in the course of developing the program attention
was drawn to areas where some modification to the evaluation procedure was
needed and this was carried out. The program was written at the Institut
für Thermodynamik und Wärmetechnik aided by valuable comments and
criticisms from the other members of the SSTG. Many evaluation runs were
carried out by all participants and these uncovered a number of program
errors which have now been eliminated. It is, however, always possible that
other errors remain hidden. The author would be grateful to hear from any
user who uncovers such errors in the future.

2. EQUATIONS FOR TEST EVALUATION AND ERROR ANALYSIS

2.1. Calculation of the time required for a heat store to reach steady state after a step change in inlet temperature (subroutine STEADY)

The equations are based on the work carried out under Subtask 2 (see [1], Part B Chapter 4).

For the estimation of the minimum time $t_{ss,min}$ required for a heat store to achieve steady state, four different limiting cases were considered. These were for stores with and without heat exchangers and with mixed or stratified response during the charge/discharge.

For a __stratified store without heat exchanger__, $t_{ss,min}$ is calculated using equation (2.1). This category of store is typified by a hot water store with direct discharge. All test data show that this type of store is steady after at least 1.5 fill times.

$$t_{ss,min} = 1.5 \, t_{cf} \tag{2.1}$$

For a __stratified store with heat exchanger__, equation (2.2) is used. The derivation of $t_{st,m}$ was carried out in Subtask 2 for the fully mixed store. However, as quite a few heat stores with heat exchanger sold as 'stratified stores' behave more or less as if they are mixed, this category of store is treated as a fully mixed store. The factor 1.5 accounts for possible errors in the estimation of the $(UA)_{f,s}$-value. The quantity $(UA)_{f,s}$ is not constant over a test: its value normally decreases, especially as the steady state is approached.

$$t_{ss,min} = 1.5 \, t_{ss,m} \tag{2.2}$$

where $t_{ss,m}$ is determined from equation (2.6).

For a __fully mixed store with heat exchanger__, $t_{ss,min}$ is again calculated from equation (2.2).

For a <u>fully mixed store without heat exchanger</u> (where there is no uncertainty with respect to $(UA)_{f,s}, \varepsilon = 1$), the following equation is used:

$$t_{ss,min} = t_{ss,m} \tag{2.3}$$

After time $t_{ss,m}$ the heating power supplied to a fully mixed store (which actually becomes steady only after infinite time) is less than 1.05 times the heat loss rate of the store, i.e. the error in $(UA)_{s,a}$ due to insufficient steady state is less than 5%:

$$\dot{Q}_H = \dot{V} \rho c_p (T_i - T_e) < 1.05 (UA)_{s,a} (T_s(t_{ss,m}) - T_a)$$

$$= \dot{V} \rho c_p \varepsilon (T_i - T_s(t_{ss,m})) < 1.05 (UA)_{s,a} (T_s(t_{ss,m}) - T_a)$$

$$\tag{2.4}$$

For discharge tests down to around ambient temperature, the right hand side of equation (2.4) can become very small leading to such large values of $t_{ss,m}$ that they exceed the test time. In those cases STEP will use the default values of $(UA)_{s,a}$ supplied in the file PARAMET.DAT.

The storage temperature $T_s(t)$ can, for a fully mixed store with heat exchanger ($UA_{f,s}$ = constant), be calculated as follows:

$$T_s(t) = k_0 + (T_s(t_0) - k_0) e^{-k_2 t} \tag{2.5}$$

where

$$k_0 = \frac{\dot{V} \rho c_p \varepsilon T_i + (UA)_{s,a} T_a}{\dot{V} \rho c_p \varepsilon + (UA)_{s,a}} \quad ,$$

$$k_2 = \frac{\dot{V} \rho c_p \varepsilon + (UA)_{s,a}}{C_{s,m}} \quad ,$$

t_o is the time at which the step occurs and ε is the heat exchanger effectiveness: $\varepsilon = 1 - e^{-NTU}$.

Substituting $T_s(t)$ in equation (2.4) and rearranging it yields $t_{ss,m}$ (equation 2.6):

$$t_{ss,m} = \frac{1}{k_2} \ln \left[\frac{T_{s,1} - k_0}{k_1 - k_0} \right] \qquad (2.6)$$

where

$$k_1 = \frac{\dot{V} \rho c_p \varepsilon T_{i,2} + 1.05 \, (UA)_{s,a} T_a}{\dot{V} \rho c_p \varepsilon + 1.05 \, (UA)_{s,a}}$$

2.2. Estimation of the storage temperature

To determine $C_{s,m}$ and $(UA)_{s,a}$, the storage temperature has to be estimated. The figure needs to be accurate, particularly for steady state conditions. If $(UA)_{s,a}$ is based on the temperature difference between the store and the ambient environment, the store temperature is also needed to calculate the heat losses during the transient between two steady state conditions. As the storage temperature normally increases or decreases rapidly (75 to 80% of the step is already completed after 2 - 3 fill times), a larger error in T_s can be accepted during the transient as this will not have a great effect on the total heat loss from one steady state to the other, and thus $C_{s,m}$.

As in the Section 2.1, a distinction has been made between four types of stores.

For the <u>stratified store without heat exchanger</u>, T_s is estimated from equation (2.7) for the steady state and from equation (2.8) during the transient ($t_o < t < t_{cf}$):

$$T_s = \frac{T_i + T_e}{2} \tag{2.7}$$

$$T_s = \frac{T_{i,1} + T_{e,1}}{2} + t_H \left(\frac{T_{i,2} + T_{e,2}}{2} - \frac{T_{i,1} + T_{e,1}}{2} \right) \tag{2.8}$$

where

$$t_H = \frac{t - t_o}{t_{cf}}$$

The maximum possible error in T_s from equation (2.7) is:

$$\delta T_s = \frac{\delta T_i + \delta T_e}{2} \tag{2.9}$$

As stated above, the error in T_s during the transient can be neglected.

For a <u>stratified heat store with heat exchanger</u>, the storage temperature can be estimated from equation (2.10) and the error in T_s from equation (2.11):

$$T_s = \frac{T_i + T_e}{2} - \frac{\Delta T_{i,e}}{NTU} \tag{2.10}$$

$$\delta T_s = \frac{\delta T_i + \delta T_e}{2} + \frac{\delta \Delta T_{i,e}}{NTU} + \frac{\delta NTU \, \Delta T_{i,e}}{NTU^2} \tag{2.11}$$

For a <u>fully mixed store with heat exchanger</u>, the storage temperature is
estimated from equation (2.12) and the error from equation (2.13):

$$T_s = T_i - \frac{\Delta T_{i,e}}{1 - e^{-NTU}} \qquad (2.12)$$

$$\delta T_s = \delta T_i + \frac{\delta \Delta T_{i,e}}{1 - e^{-NTU}} + \delta NTU \; \Delta T_{i,e} \frac{e^{-NTU}}{(1 - e^{-NTU})^2} \qquad (2.13)$$

In all cases where $T_s < T_{s,1}$, T_s is set to equal T_e in the
evaluation. This is to avoid physically impossible values of the storage
temperature ($T_s < T_{s,1}$) due to underestimation of $(UA)_{f,s}$ and during
the time when the heat transfer fluid is in the heat exchanger.

For a <u>fully mixed store without heat exchanger</u> (NTU = ∞), equation (2.12)
gives:

$$T_s = T_e \qquad (2.14)$$

and

$$\delta T_s = \delta T_e \qquad (2.15)$$

2.3. <u>Determination of the overall coefficient of heat transfer between the store and the ambient environment under finite flow conditions</u>

The overall coefficient of heat transfer between the store and the ambient
environment under finite flow conditions $(UA)_{s,a,ff}$ is determined from
equation (2.16) over a time period of t_1. It is recommended that $t_1 =$
4 hrs is used for this.

$$(UA)_{s,a,ff} = \frac{1}{t_1} \cdot \int_{t_{ss}}^{t_{ss}+t_1} \frac{\dot{V} \; \rho \; c_p \; \Delta T_{i,e}}{T_s - T_a} \; dt \qquad (2.16)$$

where t_{ss} is the time after which the store has reached steady state.

The error in $UA_{s,a,ff}$ is as follows:

$$\delta(UA)_{s,a,ff} = (UA)_{s,a} \cdot \left[\frac{\delta\rho}{\rho} + \frac{\delta c_p}{c_p} + \frac{\delta\Delta T_{i,e}}{\Delta T_{i,e}} + \frac{\delta\dot{V}}{\dot{V}} + \frac{\delta T_{a,t}}{|T_s - T_a|} + \frac{\delta T_s}{|T_s - T_a|} \right]$$

$$(2.17)$$

Relative errors in the density and specific heat capacity of water are neglected as well as the error in the measurement of the time t_1. Errors in inlet temperature, temperature difference across the store and flow rate depend on the individual test facility.

In the error of the ambient temperature T_a, not only the measurement error has to be considered but also possible fluctuations in the ambient temperature. This is because these also lead to uncertainties in $UA_{s,a}$. Obviously, fluctuations of a few minutes will not influence the result, as $(UA)_{s,a,ff}$ is calculated as mean value over 4 hours. If, however, there is a drift in ambient temperature, taking the mean ambient temperature will lead to a systematic error in the result because sufficient steady state will not have been reached.

In [1], Part B Chapter 5, the influence of non constant ambient temperature is shown. It is not possible nor worthwhile to try to include all possible changes in the error analysis. It is better to specify certain requirements for maximum allowable changes in T_a. For the remaining fluctuations, a reasonable estimation has to be applied which:

- involves an easy and general calculation, and

- weights long-term drift more strongly than short term fluctuations.

Both these requirements are met by the standard deviation in the ambient temperature. Therefore, the additional error is estimated using the standard deviation in T_a. The total uncertainty in T_a can then be calculated as follows:

$$\delta T_{a,t} = \delta T_a + \sigma(T_a) \qquad\qquad (2.18)$$

2.4. Determination of the measured heat content of the store and heat storage capacity

The rate of change of the heat content of the store is calculated from:

$$\dot{Q}_s = \dot{V}\, \rho\, c_p\, \Delta T_{i,e} - (UA)_{s,a,ff}\, (T_s - T_a) \qquad (2.19)$$

The heat charged or discharged until time t' is obtained by integrating equation (2.19) as follows:

$$Q_{s,m}(t') = \int_{t_o}^{t'} \dot{Q}_s\, dt \qquad (2.20)$$

The measured storage capacity $C_{s,m}$ is the heat charged or discharged from one steady state to the other divided by the change in storage temperature:

$$C_{s,m} = \frac{1}{T_{s,2} - T_{s,1}} \int_{t_o}^{t_{ss,min}} \dot{Q}_s\, dt = \frac{Q_{s,m}\, (t_{ss,min})}{T_{s,2} - T_{s,1}} \qquad (2.21)$$

Here, all heat supplied to the store is counted positive and all heat extracted from the store or lost to the ambient environment is counted negative.

The error in the rate of change of heat content becomes:

$$\delta \dot{Q}_s = \left[\ (\delta \dot{Q}_{s,1})^2 + (\delta \dot{Q}_{s,2})^2 \right]^{0.5} \qquad (2.22)$$

where

$$\delta \dot{Q}_{s,1} = \dot{Q}_s \left(\frac{\delta \rho}{\rho} + \frac{\delta c_p}{c_p} + \left| \frac{\delta \Delta T_{i,e}}{\Delta T_{i,e}} \right| + \frac{\delta \dot{V}}{\dot{V}} \right)$$

and

$$\delta \dot{Q}_{s,2} = \delta(UA)_{s,a,ff} (T_s - T_a) + (UA)_{s,a,ff} (\delta T_s + \delta T_a)$$

If $(UA)_{s,a,ff}$ has been determined from the steady state condition after or before a step response test, it can be assumed that the error in the heat loss and the heat input to the store under steady state are about the same. Hence the error in the net heat input to the store is only determined by the error in the heat loss of the store. In this case one of the two factors in equation (2.22) can be neglected and in STEP $\delta \dot{Q}_{s,2}$ is set to zero for this situation.

The error in the net heat input to the store becomes:

$$\delta Q_{s,m}(t') = \int_{t_0}^{t'} \delta \dot{Q}_s \, dt \qquad (2.23)$$

The error in the measured heat storage capacity is determined from:

$$\delta C_{s,m} = C_{s,m} \cdot \left(\left| \frac{\delta Q_{s,m}}{Q_{s,m}} \right| + \left| \frac{\delta T_{s,1}}{T_{s,2} - T_{s,1}} \right| + \left| \frac{\delta T_{s,2}}{T_{s,2} - T_{s,1}} \right| \right) .$$

$$(2.24)$$

2.5. Determination of the heat storage efficiency for charging or discharging

The storage efficiency for charging is the ratio of the net heat supplied to the store until time t and the total net heat input between two steady state conditions:

$$\eta_{s,c}(t) = \frac{Q_{s,m}(t)}{Q_{s,m}(t_{ss})} \qquad (2.25)$$

The error in $\eta_{s,c}$ is:

$$\delta\eta_{s,c}(t) = \eta_{s,c}(t) \cdot \left(\left| \frac{\delta Q_{s,m}(t)}{Q_{s,m}(t)} \right| + \left| \frac{\delta Q_{s,m}}{Q_{s,m}} \right| \right) \qquad (2.26)$$

where $\delta Q_{s,m}(t)$ and $\delta Q_{s,m}$ are determined with equation (2.23).

2.6. Determination of the relative heat loss and the overall coefficient of heat transfer between the store and the ambient environment during stand-by

The relative heat loss during stand-by can be determined by either recharging the store to the condition it was in before the stand-by or by discharging the store to another steady state. For the first case, the heat loss $Q_{l,sb}$ is determined by:

$$Q_{l,sb} = \int_{t_{sb,end}}^{t_{sb,end}+t_{ss}} \dot{Q}_s \, dt \qquad (2.27)$$

where $t_{sb,end}$ is the time at which the stand-by ends and t_{ss} is the time required in the recharge for the store to reach a steady state. Equation (2.27) is identical to equation (2.20) and hence the error in the heat loss can be determined from equation (2.23).

For the second case (when the store is discharged to another steady state), the heat loss is determined by:

$$Q_{1,sb} = C_{s,m} \cdot (T_{s,1} - T_s(t_{sb,end} + t_{ss})) - \int_{t_{sb,end}}^{t_{sb,end} + t_{ss}} \dot{Q}_s \, dt$$

(2.28)

The error in $Q_{1,sb}$ then becomes:

$$\delta Q_{1,sb} = \delta Q_{s,m} + \int_{t_{sb,end}}^{t_{sb,end} + t_{ss}} \delta \dot{Q}_s \, dt$$

(2.29)

The relative heat loss for both cases is calculated from:

$$dQ_1 = \frac{Q_{1,sb}}{C_{s,m} \, (T_{s,1} - T_{a,sb})}$$

(2.30)

The error in the relative heat loss is obtained from:

$$\delta dQ_1 = dQ_1 \cdot \left(\frac{\delta Q_{1,sb}}{Q_{1,sb}} + \frac{\delta Q_{s,m}}{Q_{s,m}} \right)$$

(2.31)

The overall coefficient of heat transfer between the store and the ambient environment during stand-by is calculated from:

$$(UA)_{s,a,sb} = - \frac{C_{s,m}}{t_{sb}} \ln \left(\frac{C_{s,m} \, (T_{s,1} - T_{a,sb}) - Q_{1,sb}}{C_{s,m} \, (T_{s,1} - T_{a,sb})} \right)$$

(2.32)

and the error in $(UA)_{s,a,sb}$ from:

$$\delta UA_{s,a} = (\frac{\delta Q_{1,sb}}{Q_{1,sb}} + \frac{\delta Q_{s,m}}{Q_{s,m}}) \frac{Q_{1,sb}}{(C_{s,m}(T_{s,1}-T_{a,sb})-Q_{1,sb}) \ln(1- \frac{Q_{1,sb}}{C_{s,m}(T_{s,1}-T_{a,sb})})}$$

$$+ \frac{\delta C_{s,m}}{C_{s,m}} \qquad\qquad (2.33)$$

2.7. Determination of the heat loss of the top of the store and the overall coefficient of heat transfer between the upper part of the store and the ambient environment

The heat loss of the top half of a store can be determined by either discharging or reheating the partly discharged store after a stand-by period. The initial condition for the stand-by is achieved by heating the store first to steady state and then discharging about half the store volume. For the reheat case, the heat loss Q_1 is determined by:

$$Q_{1,top} = \int_{t_{sb,end}}^{t_{sb,end}+t_{ss}} \dot{Q}_s \, dt \; - Q_{d,1} \qquad\qquad (2.34)$$

where

$$Q_{d,1} = \int_{t_{d,ini}}^{t_{d,end}} \dot{Q}_s \, dt$$

and $t_{d,end} - t_{d,ini} = 0.5\, t_{df}$ and $t_{d,end} = t_{sb,ini}$.

For the discharge case, the heat loss is determined from equation (2.35):

$$Q_{1,top} = C_{s,m} \; (T_{s,1} - T_s(t_{sb,end}+t_{ss})) - Q_{d,1} - Q_{d,2} \qquad (2.35)$$

where

$$Q_{d,2} = \int_{t_{sb,end}}^{t_{sb,end}+t_{ss}} \dot{Q}_s \, dt$$

The overall coefficient of heat transfer between the upper part of the store and the ambient environment can now be calculated for both test types using:

$$(UA)_{s,a,top} = - \frac{C_{s,m,top}}{t_{sb}} \ln \left[\frac{C_{s,m}(T_{s,1} - T_{a,sb}) - Q_{1,top} - Q_{d,1}}{C_{s,m}(T_{s,1} - T_{a,sb}) - Q_{d,1}} \right]$$

(2.36)

where $C_{s,m,top} = C_{s,m} - Q_{d,1}/(T_{s,1} - T_a)$

3. USER GUIDE

3.1. Introduction

To speed up compiling and linking, the whole program has been broken down
into the following five sections. These are also used in the diskette
supplied with this report.

The main program STEP
The section   SUBIN containing subroutines INPUT and MENUE
The section   SUBCAL containing subroutines STEMP, HLRATE, QMSC, ETA,
                  STEPQ, STEADY, NUMBER, QSIMP, SDEV
The section   SUBOUT containing subroutines OUTPUT, TABLE and PLOTF
The section   SUBPRP containing functions FCP and FRHO

3.2. Subroutine MENUE

Subroutine MENUE reads a set of parameters from a file called PARAMET.DAT
and displays them on the computer screen. A printout of MENUE is shown in
Tables 3.1 and 3.2; an example of the file PARAMET.DAT is shown in Table
3.3.

On the first output page, all parameters have a number and the typing of
this number enables this parameter to be changed. The computer will ask for
a new value. After this has been typed in correctly, the first page is
updated, showing the modified value. The updated values may be written into
the file PARAMET.DAT by typing in 1 at the end of the second page of the
screen output.

```
┌─────────────────────────────────────────────────────────────────────┐
│                 SOLAR STORAGE TESTING GROUP                          │
│                                                                       │
│          FORTRAN 77 Test Evaluation Program  STEP                    │
│          Version 2.2, ITW, University of Stuttgart 1989              │
│                                                                       │
│    1 Test Type: HLZFDC            FILES:    2 Data from : EXR211.DAT  │
│                                             3 Results   : EXR211.TES  │
│                                             4 Plotfile  : NOFILE      │
│                                                                       │
│    6 Cs,m    :   3897.2 ±   90.80 kJ/K                                │
│    7 Cs,t    :   4187.0 ±  200.00 kJ/K       8 time to reach          │
│    9 (UA)s,a:      4.0 ±     .50  W/K          steady state :    .0 h │
│                                                                       │
│    Heat Exchanger:                                                    │
│                                             10 CHARGE       : HEXCH   │
│   12  UAhx :  500.0 ±  100.0 W/K            11 DISCHARGE    :NOHEXCH  │
│                                                                       │
│    Response                                 13 CHARGE       : MIXED   │
│                                             14 DISCHARGE    : STRAT   │
│                                                                       │
│   15 Unit of flow rate : L/H                                         │
│   II  Changes ? enter code nr.   (no change = 0)                     │
└─────────────────────────────────────────────────────────────────────┘
```

Table 3.1: Screen output of subroutine MENUE (first page).

```
┌─────────────────────────────────────────────────────────┐
│  Heat transfer fluid: Enter 0 for water                 │
│                             1 for water/glykol          │
│                                                          │
│          16 charge    loop :   0                        │
│          17 discharge loop :   0                        │
│                                                          │
│                                                          │
│    Maximum error              systematic                │
│          Ti                   ±   .10 K                 │
│          Te                   ±   .10 K                 │
│         dTi,e                 ±   .03 K                 │
│          Ta                   ±   .10 K                 │
│    Rélative Error                                       │
│          Vdot                 ±  .010                   │
│          Density              ±  .000                   │
│          Specific Heat Capacity ± .000                  │
│  SAVE PARAMETERS ? - YES enter 1                        │
└─────────────────────────────────────────────────────────┘
```

Table 3.2: Screen output of subroutine MENUE (second page).

The different parameters are:

1. Test type: HLFF   - heat loss test at finite flow rate (no step)

                CHARGE   - temperature step response charge test

                DCHARGE - temperature step response discharge test

                HLZFRH   - relative heat loss test, reheat mode

                HLZFDC   - relative heat loss test, discharge mode

                HLTOPR   - top heat loss test, reheat mode

                HLTOPD   - top heat loss test, discharge mode

2 - 4.   Names of files with measured data, output, plot data. If
                the name of the plotfile is NOFILE, no file will be
                created.

6. $C_{s,m}$ An estimation of the measured storage capacity should be
                given here. It is needed for the relative heat loss
                tests. This default value will be overwritten during
                program execution of charge and discharge tests (but is
                not written to PARAMET.DAT).

7. $C_{s,t}$ As in 6. $C_{s,t}$ is only used for the output.

8. t_{st} If a value > 0 is put in here, this will be used and
                subroutine STEADY be overruled. If the value is greater
                than the test time, this will cause the program to use the
                default value of $(UA)_{s,a}$.

9. $(UA)_{s,a}$ Default value for $(UA)_{s,a}$, which will be used if a heat
                loss coefficient of heat transfer cannot be calculated in
                the program itself (because steady state is not reached or
                there is too low a storage temperature). In combination
                with 8, the $(UA)_{s,a}$-value calculated in the program can
                be overruled and the default value be used.

10 - 12.   The heat exchanger with the UA_{hx} (i.e. $(UA)_{f,s}$)-value
                specified is considered in the operation condition
                specified. For charge or discharge either HEXCH or NOHEXCH
                can be chosen. The program switches automatically,
                controlled by the status variable.

13 - 14.   Same principle as 10 - 11, but with regard to the
                response.

15.   Four different units of the volume flow rate can be
                handled in the program automatically: l/h, l/s, m³/h and
                m³/s.

16 - 17.   Selection of the heat transfer medium in charge and
                        discharge loop. Properties for water and a water/glycol
                        mixture (50% / 50%) are available in the functions FCP and
                        FRHO. The concentration can be changed in these functions.

The second page of the screen output displays the error parameters but does
not allow for changes: these have to be made directly in the file
PARAMET.DAT.

The file PARAMET.DAT (Table 3.3) contains the following parameters (format
in brackets):

Parameter	Variable Name	Format
Test type	TYPE	(Character. A8)
Heat transfer medium, charge/discharge loop	MEDCL,MEDDCL	(Integer.2I2)
Heat exchanger in charge/discharge	CHX,DHX	(Character,2A8)
Response during charge/discharge	CRESP,DRESP	(Character,2A6)
time required to reach steady state	THRST	(Real, F6.1)
Unit of flow rate	VUNIT	(Character, A4)
Filenames	DAINP,DAOUT	(Character,3A12)
	DAOUTP	
$C_{s,m}$, $C_{s,t}$, $UA_{s,a}$, UA_{hx}	MSCD,TSCD	(Real, 4F7.1)
	UASA,UAHX	
$\delta C_{s,m}$, $\delta C_{s,t}$,	ERMSCD,ERUAHX,	
$\delta UA_{s,a}$, δUA_{hx}	ERUASA,ERTSCD	(Real, 4F7.1)
Start of heat loss test	NDS,IHS,IMS	(Integer, 3I3)
Errors in T_i, T_e, $dT_{i,e}$, T_a,	ERTI,ERTE	(Real, 4F6.2)
	ERDT,ERTA	
Relative Errors of $\dot{V}$, ρ, c_p	RERV,RERRHO,RERCP	(Real, 3F6.3)

```
HLTOPR
 0 0
HEXCH    NOHEXCH
MIXED STRAT   100.0
L/H
S7E3.HEI    S7E3.RES    NOFILE
 3660.0 4187.0   4.20 700.00
 100.00 200.00   0.50 200.00
 12 12 12
  0.10  0.10  0.02  0.10 0.010 0.000 0.000
```

Table 3.3: File PARAMET.DAT

3.3. Subroutine INPUT

Subroutine INPUT reads the experimental data from the input file. The
following order and formats have to be used:

Year, month, day, hour, minute, second, T_i , T_e , $dT_{i,e}$, V , T_a , TZ , IST, P_{aux}

14 , I2 , I2 , I2 , I2 , I2 , F6.2, F6.2, F7.3 , F7.1, F6.2 , F6.2, I3 , F8.1

The variable TZ should contain the values of a zeroed channel in the data
acquisition. STEP calculates a standard deviation of a zero offset and
prints the results in the output file (see Section 3.5). The measured data
are stored in arrays (maximum 2500 data points!). Before storing the data,
the flow rate is converted to the unit m^3/s and the time in hours relative
to the start of the test. A check is performed on the time order of the
data and if $t(J) \geq t(J+1)$ a message is printed on the screen: 'Inconsistent
time, data point N'. If an error during input occurs, the message 'Read
error in data point No.: N' is printed, so that the relevant data record
can be found easily.

3.4. Program structure

With the information supplied in the MENUE and in the data file, the
program runs automatically. In the evaluation of step charge tests, the
program searches for a gradient in the inlet temperature greater than 100
K/h; the data point before the occurance of this gradient is assumed to be
the start of the step (NSTART). From NSTART backwards for 4 hours, the
initial steady state condition is calculated. If 4 hours is not availbale,
3, 2 or 1 hour is used.
The final steady state condition starts after the time t_{SS}. This can
either be chosen in MENUE or calculated by the program (if t_{SS} = 0 in the
MENUE).
The evaluation of relative heat loss test is controlled by the status
variable IST which is 1 during charge, 0 during stand-by and -1 during
discharge.
For the evaluation of pure heat loss tests, the start time has to be given
in MENUE.

3.5. Output

The standard output is written in the file specified in MENUE. An example is given at the end of this section. The output file contains the MENUE, the results and a table with the hourly mean values of the test.

SOLAR STORAGE TESTING GROUP

FORTRAN 77 Test Evaluation Program STEP
Version 2.2, ITW, University of Stuttgart 1989

1 Test Type: CHARGE   FILES: 2 Data from : a12.dat
                                           3 Results   : A11.OUT
                                           4 Plotfile   : NOFILE

6 Cs,m :   0.8875 ± 0.1100 MJ/K
7 Cs,t :   0.9250 ± 0.1000 MJ/K   8 time to reach
9 (UA)s,a: 2.50 ±   0.30 W/K   steady state :   22.9 h

Heat Exchanger:
                                           10 CHARGE   : HEXCH
12 UAhx : 120.0 ±   30.0 W/K   11 DISCHARGE   :NOHEXCH

Response   13 CHARGE   : MIXED
                                           14 DISCHARGE   : STRAT

Heat Transfer Fluid: Heating Loop: Water
                        Cooling Loop: Water

    Maximum error   systematic
        Ti   ±   0.10 K
        Te   ±   0.10 K
        dTi,e   ±   0.05 K
        Ta   ±   0.60 K
    Relative Error
        Vdot   ± 0.004
        Density   ± 0.005
        Specific Heat Capacity ± 0.005

    Zero-offset:
        Mean Value   0.0000 Mikrovolt
        Standard Deviation   0.0000 Mikrovolt

DATA SHEETS FOR
TEMPERATURE STEP RESPONSE CHARGE   TEST

Reference number:_____ Test number:_____

1. test conditions
   - temperature step   Ti =   51.4 -   71.5 °C
   - average flow rate over tcf: Vdot = 0.5453E-04 M3/S

2. initial conditions
   - heat transfer fluid inlet temperature Ti =   51.4 °C
   - temperature difference across store dTi,e=   0.33 K
   - ambient air temperature   Ta =   19.5 ± 0.7 °C
   - flow rate   Vdot = 0.5459E-04 M3/S
   - heat loss capacity rate   UAs,a =   2.37 ± 0.47 W/K
   - averaging time   ta =   4.01 hours

3. final conditions
   - heat transfer fluid inlet temperature Ti =   71.5 °C
   - temperature difference across store dTi,e=   0.55 K
   - ambient air temperature   Ta =   19.8 ± 0.6 °C
   - flow rate   Vdot = 0.5453E-04 M3/S
   - heat loss capacity rate   UAs,a =   2.45 ± 0.31 W/K
   - averaging time   ta =   5.08 hours

4. results
   Qs,m   =   16.943 ± 1.331 MJ   Qs,m'=   0.8875 ± 0.1100 MJ/K
   Qs,t   =   17.658 ± 2.711 MJ   Qs,t'=   0.9250 ± 0.1000 MJ/K
   tcf   =   1.050   HOURS
   ETA05TCF   =   0.310 ± 0.030
   ETA1TCF   =   0.530 ± 0.052
   ETA2TCF   =   0.773 ± 0.078
   ETA1H   =   0.520 ± 0.051
   t95   =   0.025   HOURS

from STEP to the end of the test:
   STANDARD DEVIATION OF TI2 =   0.051   K
   STANDARD DEVIATION OF VDOT =   0.909   %
   STANDARD DEVIATION OF Ta   =   0.199   K

5. MEASURED DATA

TABLE 5.

				average values over time steps			of 1 hour					at end of time step	
day	time interv.	T_i	SDEVTI	$dT_{i,e}$	T_e	T_a	Vdot	SDEVVD	Qdot	UAsa	Qldot	tau	eta
		°C	K	K	°C	°C	m3/s	%	W	W/K	W		
18	1	51.4	0.000	0.33	51.1	19.4	0.546E-04	0.00	74.6	2.37	75.2	0.00	0.00
18	2	51.4	0.000	0.33	51.1	19.4	0.545E-04	0.00	75.0	2.37	75.2	0.00	0.00
18	3	51.4	0.000	0.34	51.1	19.5	0.546E-04	0.00	75.6	2.37	74.8	0.00	0.00
18	4	51.4	0.000	0.34	51.1	19.6	0.546E-04	0.00	75.6	2.37	74.7	0.00	0.00
18	5	69.3	0.070	10.25	59.1	19.6	0.547E-04	1.18	2296.7	2.40	91.5	0.94	0.51
18	6	71.5	0.060	5.46	66.0	19.7	0.545E-04	1.09	1217.1	2.40	92.9	1.89	0.74
18	7	71.5	0.058	2.85	68.6	19.7	0.545E-04	1.09	635.4	2.43	109.0	2.84	0.86
18	8	71.5	0.056	1.70	69.8	19.8	0.545E-04	1.06	378.1	2.44	116.1	3.79	0.92
18	9	71.5	0.055	1.18	70.3	19.9	0.546E-04	1.02	263.9	2.44	119.2	4.71	0.95
18	10	71.5	0.055	0.91	70.6	19.9	0.545E-04	1.01	203.9	2.45	120.7	5.70	0.97
18	11	71.5	0.055	0.76	70.7	19.9	0.546E-04	1.01	170.8	2.45	121.8	6.65	0.99
18	12	71.5	0.054	0.69	70.8	19.9	0.543E-04	1.01	153.6	2.45	122.3	7.60	0.99
18	13	71.5	0.054	0.65	70.8	19.9	0.546E-04	1.00	144.6	2.45	122.5	8.56	1.00
19	14	71.5	0.054	0.62	70.9	19.9	0.544E-04	1.00	137.8	2.45	122.6	9.51	1.00
19	15	71.5	0.054	0.61	70.9	19.9	0.545E-04	0.99	135.4	2.45	122.7	10.46	1.00
19	16	71.5	0.054	0.59	70.9	19.9	0.546E-04	0.98	131.8	2.45	122.8	11.41	1.01
19	17	71.5	0.054	0.58	70.9	20.0	0.544E-04	0.98	128.1	2.45	122.9	12.37	1.01
19	18	71.5	0.054	0.57	70.9	19.9	0.547E-04	0.97	128.0	2.45	123.0	13.32	1.01
19	19	71.5	0.054	0.57	70.9	19.9	0.544E-04	0.97	126.1	2.45	123.1	14.27	1.01
19	20	71.5	0.054	0.57	70.9	19.9	0.546E-04	0.96	126.6	2.45	123.0	15.22	1.01
19	21	71.5	0.054	0.56	70.9	20.0	0.545E-04	0.96	125.6	2.45	122.9	16.18	1.01
19	22	71.5	0.054	0.56	70.9	20.0	0.546E-04	0.96	125.2	2.45	122.9	17.13	1.01
19	23	71.5	0.054	0.56	70.9	20.0	0.544E-04	0.95	124.6	2.45	122.9	18.08	1.01
19	24	71.5	0.054	0.56	70.9	20.0	0.546E-04	0.94	124.6	2.45	123.0	19.03	1.02
19	25	71.5	0.053	0.55	70.9	20.0	0.545E-04	0.94	123.0	2.45	123.0	19.99	1.02
19	26	71.5	0.053	0.55	70.9	19.9	0.547E-04	0.93	123.4	2.45	123.1	20.94	1.02
19	27	71.5	0.053	0.55	70.9	19.9	0.546E-04	0.92	122.2	2.45	123.2	21.89	1.02
19	28	71.5	0.052	0.55	70.9	19.9	0.545E-04	0.92	123.2	2.45	123.2	22.84	1.02
19	29	71.5	0.052	0.55	70.9	19.9	0.545E-04	0.91	123.2	2.45	123.2	23.80	1.02
19	30	71.5	0.052	0.55	70.9	19.9	0.545E-04	0.91	123.6	2.45	123.3	24.75	1.02
19	31	71.5	0.051	0.55	70.9	19.8	0.547E-04	0.91	123.8	2.45	123.3	25.70	1.02

REFERENCES

[1] Visser, H. and H.A.L. van Dijk (ed.)
    Final Report on the Activities of the Solar Storage Testing Group.
    Commission of the European Communities, EUR 13119.

APPENDIX D:

THE PROTOTYPE 4PORT STORE MODEL AND USAGE PROGRAMS FOR IDENTIFICATION
AND VALIDATION*)

CONTENTS

*) Chapters 1, 2 and 3 are a summary based on the final report on the
   model 4PORT by R.M. Marshall and C.L.W. Li of the Division of Mechanical
   Engineering and Energy Studies, University of Wales College of Cardiff,
   UK, see [2], Part B Chapter 1.

1. INTRODUCTION

The objective of the development of a general storage model is to derive a
model for use in the long term system performance estimate whose parameters
are derived from short term testing. It is the requirement of a general
model that dictates the need for a computer code together with system
identification techniques in order to cater for cases where simple analytic
models and simple experiments are no longer valid.

In particular, there occur storage devices with a heat exchanger with a
large thermal capacity and hence a large residence time. The store and/or
the heat exchanger may operate in a stratified manner. In short, it is not
always possible to assume a steady state heat exchanger equation coupled to
a fully mixed store. In addition, the input test sequence may be imperfect
so that analytic theories based on ideal boundary conditions cannot be
realised. The governing equations may also be temperature, flow rate and
even time dependent.

These considerations, then, lead to the choice of a numerical model using
finite differences. The key parameter to be identified from step response
testing is the heat transfer fluid to storage material heat transfer
coefficient, denoted as $(UA)_{f,s}$ below.

The general model, called 4PORT, is described in Chapter 2. This chapter
includes a presentation about setting up the model.

Chapter 3 describes how the model can be used for validation against the
test results of the dynamic test, using the usage program DYNAMI and for
identification of $(UA)_{f,s}$ using the usage program IDENUA.

Chapter 4 introduces the background and use of the usage program MRQ-4PORT,
a multiparameter identification method.

2. THE PROTOTYPE 4PORT MODEL

2.1. Model description

2.1.1. The model

The prototype model is shown schematically in Figure 2.1. The storage
material is divided into NS storage zones. The heat exchanger fluid is
divided into a fewer number of nodes, NF. We allow NDEAD nodes to denote
the "dead volume" in the storage material.

The equation describing a fluid node is given by:

$$\frac{Mc_{p,f}}{NF} \frac{\partial T_{f,i}}{\partial t} = \dot{m}c_{p,f}(T_{f,i-1} - T_{f,i}) + \binom{0}{1} \frac{(UA)_{f,s}}{NF} (T_{s,j} - T_{f,i})$$

$$+ \frac{(UA)_{f,a}}{NF} (T_a - T_{f,i}) \tag{2.1}$$

The storage node equation is:

$$\frac{Mc_{p,s}}{NS} \frac{\partial T_{s,i}}{\partial t} = \dot{m}c_{p,s}(T_{s,j+1} - T_{s,j}) + \frac{(UA)_{s,a}}{NS} (T_a - T_{s,j})$$

$$+ \binom{0}{1} \frac{(UA)_{f,s}}{NF} (T_{f,i} - T_{s,j}) + (UA)_{s,s}\Big|_{j+\frac{1}{2}}(T_{s,j+1} - T_{s,j})$$

$$+ (UA)_{s,s}\Big|_{j-\frac{1}{2}}(T_{s,j-1} - T_{s,j}) \tag{2.2}$$

Here T_f, T_s, and T_a denote the heat exchanger fluid, store, and
ambient temperatures respectively.

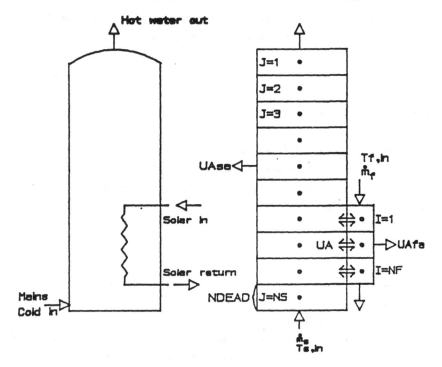

Figure 2.1: The prototype 4PORT store model.

The charge and discharge model capacity rates are denoted as $\dot{m}c_{p,f}$ and $\dot{m}c_{p,s}$ respectively; $(UA)_{f,a}$ denotes the overall coefficient of heat transfer between fluid (f) and ambient (a) and $(UA)_{s,a}$ from the store (s) to ambient. $(UA)_{s,s}$ denotes the internodal conduction between adjacent storage nodes. This term is an "effective" value in that it accounts for three dimensional effects and turbulent jet mixing effects especially on discharge. In essence each nodal equation states that the rate of exchange of internal energy is equal to the energy transported in/out of the cell by forced convection, plus the energy transported in/out at the heat exchanger-store material interface, minus the energy lost to the ambient.

Of these, the {0 1} denotes the binary switch function to include the term
whenever a fluid node is in contact with the storage node namely,

$$J = I + (NS-NDEAD) - NF \text{ for } I = 1 \text{ to } NF \qquad (2.3)$$

$$I = J + NF - (NS-NDEAD) \text{ (positive I only)} \qquad (2.4)$$

The coupled equations (2.1) and (2.2) are solved numerically. An explicit
scheme is used when the model is used for identification and short term
validation while a semi-implicit solution is used for the estimate of the
long term system performance.

2.1.2. Mixing

With a heat exchanger located near the bottom of the store heat is
transferred upwards by means of natural convection. This process depends on
several factors of which the primary one is the temperature difference
between the heat exchanger surface temperature and that of the surrounding
fluid. The precise modelling of the convective transfer is, in general, far
too time consuming for use in, say, an hour-by-hour simulation. Also, the
process time scale is relatively quick. That is, hotter fluid below cooler
is an unstable process so that mixing begins. In our model we will replace
the true mixing transport with a much simplified version in order to save
computing time. At the end of each computational step the temperatures in
the storage nodes are scanned from top to bottom, i.e. from J = 1 to J =
NS, in order to find the maximum temperature. If this node is not 1 we
"mix" i.e. average the storage temperatures from 1 down to the cell. Again
all the cells are scanned for the maximum value and further mixing is
performed if necessary. In this way a stably stratified store, one with
hotter fluid above colder fluid, is always achieved.
Provided that the true time-scale of the mixing process is short (say, less
than five minutes or of that order), this simplified mixing process can be
expected to give good agreement with our experimental results.

2.1.3. The numerical solution

Equations (2.1) and (2.2) represent an approximation to the fundamental
energy equation for each material.
In [2], Part B Chapter 1 the consequences are explained, in particular the
occurence of the socalled "artificial" or "numerical" diffusion. Numerical
diffusion can only be avoided either by a time step size which matches
exactly the fill time of each store segment in the model ($f = 1$) or by a
large number of nodes. The former option may, however, conflict with
requirements with respect to a stable solution.

2.2. Use of the model 4PORT

2.2.1. Usage programs

The model given by equations (2.1) and (2.2) has been coded in FORTRAN 77
as a stand alone subroutine called FOURPO. It must be called by a MAIN
routine whose job is to
a) set up the model parameters,
b) set the initial fluid and store temperatures,
c) call the subroutine FOURPO with suitable boundary conditions,
d) process the results and
e) terminate the simulation.
Two main usage programs have been developed by the Energy Studies Unit of
the University in Cardiff, UK:
a) IDENUA for the identification of $(UA)_{f,s}$ using step response data and
b) DYNAMI for the validation of the model by means of dynamic testing.
By the TNO Institute of Applied Physics in Delft, The Netherlands a third
usage program was developed, MRQ-4PORT for more general parameter
identification purposes. All three application programs make use of the
same storage model, 4PORT. Therefore, the first task is to set up the
model.

2.2.2. <u>Translating the store geometry into model parameters</u>

The first task is to specify the model parameters: M_f, M_s, $(UA)_{f,s}$,
$(UA)_{f,a}$, $(UA)_{s,a}$, and $(UA)_{s,s}$. In the current version of 4PORT both
the heat transfer fluid and storage material are assumed to be water so
that the specific heat capacity is known, about 4185 J/kg K.
The model parameters may have been determined from processing step response
data using, for example, STEP version 2.2 or known from measurements or
estimated from manufacturer's data.

2.2.3. M_f and M_s

From STEP one obtains the total thermal capacity of the store, $C_{s,m}$. If
M_f is known then the equivalent M_s can be found from

$$M_s = (C_{s,m} - M_f c_{p,f})/c_{p,s} \qquad\qquad (2.5)$$

Note that it is possible that the measured $(C_{s,m})$ and theoretical
$(C_{s,t})$ thermal capacities differ. In this case the equivalent M_s will
differ from that computed, say, from the known volume of the store and
density of the storage material. In most cases the measured capacity ought
to be used in preference to the theoretical and M_s should be determined
from equation (2.5).

2.2.4. <u>Relationship between $(UA)_{f,s}^{model}$ and $(UA)_{f,s}$</u>

The determination of the key model parameter $(UA)_{f,s}^{model}$ is the primary
reason for using the model 4PORT in conjunction with a suitable
identification package, e.g. the Marquardt algorithm, see Press et al.
(1987) [1], p. 523. Strictly speaking $(UA)_{f,s}^{model}$ is only a model parameter
and its value depends on the number of nodes chosen for the heat transfer
fluid, NF.

The link to the physical parameter (call it $(UA)_{f,s}$) governing the heat transfer between the fluid and the storage material is given by the lemma:

In the limit as NF and, by necessity, NS tends to infinity, the model parameter $(UA)_{f,s}^{model}$ will approach the 'physical' $(UA)_{f,s}$.

For finite and especially smaller values of NF one must use a larger model parameter $(UA)_{f,s}$ in order that the computed response be correct. Refer to the section below on identification for interpreting the $(UA)_{f,s}$-value.

2.2.5. $(UA)_{f,a}$ and $(UA)_{s,a}$

These two parameters govern the total heat loss from the store. From the data analysis of STEP or from the manufacturer's data about the type and thickness of the insulation, the total loss rate [W/K] from the store is estimated. If the store is equipped with a "mantle" type heat exchanger shell of area A_{mantle} around part of the store, then the heat exchanger fluid can lose heat directly through the insulation to the ambient environment. Thus the total heat loss, $(UA)_s$, must be apportioned in terms of the surface area of the mantle, if present, to the total area of the store, namely:

$$(UA)_{f,a} = (UA)_s * A_{mantle}/A_s \qquad (2.6)$$

$$(UA)_{s,a} = (UA)_s * (A_s - A_{mantle})/A_s \qquad (2.7)$$

2.2.6. $(UA)_{s,s}$

The internodal conduction term under normal circumstances is a negligible quantity. If the tank is idealized to have a volume, V_s [m³], and a height, L_s [m], then the equivalent cross sectional area is simply V_s/L_s. The nominal internodal conduction $U_{s,s}$-value is given by

$$U_{s,s} = \lambda_s/(L_s/NS) \qquad (2.8)$$

where λ_s is the thermal conductivity of the storage medium.

The nominal internodal conduction coefficient is simply.

$$(UA)_{s,s} = \lambda_s \ NS \ V_s/(L_s^2) \tag{2.9}$$

However, due to turbulent mixing of the incoming jet of fluid upon direct discharge and due to thermal conduction in the container wall a one dimensional plug flow model is no longer valid. The effect can be approximated by using a $(UA)_{s,s}$-value which is some two to four orders of magnitude larger than the nominal $(UA)_{s,s}$ given by equation (2.9). The effects of turbulent mixing are seen as a "smeared" discharge profile whereby a sharp thermocline no longer emerges from the top of the store. The model 4PORT can be used together with an identification routine, e.g. the Marquardt algorithm, provided that a large number of nodes (say, 100) or even very large (say, 500) are used in the model for the parameter NS. Otherwise, the effects of the numerical smearing associated with a small number of nodes, NS, will mask the actual turbulent mixing contribution; see in Part B Sections 7.3.3 and 7.3.4 how this is accounted for in the instructions.

2.3. Choice and placement of the nodes

The choice of the number of nodes to represent the heat exchanger, the storage material and the dead zone reflects the use of the model and the accuracy required. The placement of the nodes is also crucial for obtaining a meaningful result. It is assumed that the user of 4PORT will have sufficient experience and knowledge of the store to be modelled for the choice of the number of nodes and their placement to be straightforward. [2] gives examples taken from the SSTG testing programme.

2.3.1. In the store, NS

The choice of the number of storage nodes, NS, is made by looking at the discharge profile. If the discharge exit temperature versus time curve is mixed so that the exit temperature drops immediately after the step, then a single storage node is required, i.e. NS = 1. If the discharge profile is stratified so that a period of constant exit temperature is obtained, then a larger NS must be chosen.

See [2], Part B Chapter 1 for a more extensive discussion.
See in Part B Chapter 7 how this is accounted for in the instructions.

2.3.2. In the heat exchanger, NF

The number of heat exchanger fluid nodes is estimated by comparing the
physical vertical height of the heat exchanger (call it L_f) and the
vertical height of the store called L_s [m]. The number of fluid nodes is
then simply,

$$NF = NS \, L_f/L_s \qquad (2.10)$$

In the usual case the heat exchanger height, L_f, can be taken as the
distance between the heat exchanger inlet and outlet port. However, some
care must be taken as it will be often necessary for the user to consult
the manufacturer's drawings to establish the effective height and position
of the active portion of the heat exchanger because the ports do not always
correspond directly to the heat exchanger position.

2.3.3. In the dead zone, NDEAD

One must set the value of the number of dead zones, NDEAD. Again some care
must be taken and the manufacturer's data sheets must be examined. As
before, it is the distance below the active part of the heat exchanger
(call it L_{dead} [m]) which needs to be established from which NDEAD is
computed as,

$$NDEAD = NS \, L_{dead}/L_s \qquad (2.11)$$

2.3.4. In the auxiliary (electric) zone, NAUX

In a similar fashion the number of auxiliary nodes, NAUX, can be
estimated. For the present, only an electrical auxiliary is considered.
Denoting the vertical height as L_{elec} [m], NAUX is given by:

$$NAUX = Ns \, L_{elec}/L_s \qquad (2.12)$$

If the physical position of the electrical auxiliary is not at the top of
the store, then a horizontal auxiliary is assumed at position $L_{h,aux}$
(L-horizontal-auxiliary), and the auxiliary acts as just a single storage
node position determined in the code.

2.4. Extension for higher accuracy

In [2], Part B Chapter 1 an extensive discussion can be found on 4PORT
running in a "low resolution" versus in a "high resolution" mode, i.e. with
a low or high number of nodes.
Naturally, the computational time for a high degree of accuracy increases
sharply, roughly proportional to the square of the number of fluid nodes.
Unless extreme accuracy is required, the use of the lowest number of nodes
should be sufficient for most uses of 4PORT.

2.5. Running the simulation

The model 4PORT is run with a suitable set of initial conditions and
boundary conditions. Input temperature, T_i, mass flow rate, $\dot{m}(t)$, and
ambient temperature, $T_a(t)$, profiles are used as the boundary conditions
together, usually, with a measured temperature difference profile,
$\Delta T_{i,e}(t)$, for use in comparing the model response with that of an
experiment.
An example of an INPUT data file is given in the next chapter.

2.6. Future work

2.6.1. 6PORT

Despite the good capabilities of model 4PORT, some important areas still
need work. The first area is the extension to model an auxiliary heat
exchanger coupled to a boiler. The use of a 6PORT model would be a logical
extension of the present approach.

2.6.2. <u>Temperature, time and flow rate dependency</u>

The key model parameter, $(UA)_{f,s}$, to date has been assumed in 4PORT to be
temperature and time independent. Other models could be applied wherein a
temperature, time and flow rate dependency is assumed. Some consideration
of those models for $(UA)_{f,s}$ must be considered and would be easily
implemented in 4PORT.

2.6.3. <u>Other stores</u>

Although water storage is the most common storage medium, consideration of
other storage materials, should be made. Extending the model 4PORT to
include rock bed stores with air as the heat transfer medium and phase
change storage devices would appear to be feasible and without many changes
to the existing code.

3. THE USAGE PROGRAMS IDENUA AND DYNAMI

3.1. Identification program IDENUA

The primary use of 4PORT is for the identification of the $(UA)_{f,s}$-value
since there is no simple analytic solution which yields the value for a
store with a large heat exchanger capacity, with losses, and with a dead
zone or in the case when $(UA)_{f,s}$ is temperature dependent. Additional
parameter identification can also be done using a multilinear regression
algorithm like the Marquardt method, see chapter 4.

The $(UA)_{f,s}^{model}$-value is found by minimizing the squared error between the
measured heat exchanger exit temperature and that for the model. As noted
above, the identified $(UA)_{f,s}^{model}$ is only a model parameter and when the
number of fluid nodes, NF, is small (especially NF = 1) its value will
differ from that obtained by making internal measurements. The physical
average $(UA)_{f,s}$ can be found reliably either by running 4PORT with a
large number of fluid nodes (e.g. NF = 10, for which all of the other nodal
choices must be scaled appropriately) or by extrapolation (see equation
(2.30) in Part B). IDENUA also provides the necessary mean temperatures to
calculate c_O according to equation (2.31) (Part B).

3.2. Validation using the model 4PORT with the program DYNAMI

After identification of the model parameters using step response data, the
next task is model validation using a dynamic test sequence. The
description of the test sequence, use of the model and data analysis can be
found in Part B Chapter 5. We state here only the objective of the dynamic
test, namely to demonstrate that the identified model parameters lead to a
'correct' prediction of the transient behaviour under more general and
extreme operating conditions. Two operating conditions are considered: one
without any auxiliary heater in operation and one with the (electric)
auxiliary heater, if fitted. The criterion for judging the model is based
on the root mean squared error between measured and predicted energy flows
as well as temperature levels.

3.3. <u>Explanation and examples of parameters input file (PRGDAT) for programs DYNAMI AND IDENUA</u>

DAINP Name of file which contains the experimental data.
     Input format: a maximum of 20 characters
     Example of input --> B-D2.DAT

DAPLT Name of file which will store the simulation and
     experimental results for graph plotting.
     Note: results will not be stored if the input file
        name is 'NOFILE' (capital letters).
     Input format: a maximum of 20 characters
     Example of input --> B-D2.PLT

DARES Name of file which will store the energy balance
     results.
     Note: results will not be stored if the input file
        name is 'NOFILE' (capital letters).
     Input format: a maximum of 20 characters
     Example of input --> B-D2.RES

METHOD Choice of numerical methods.
     Input list: EXPLICIT (this will set SIGF=SIGS=0)
          or
          IMPLICIT (this will set SIGF=SIGS=0.5)
     Note: 'EXPLICIT' or 'IMPLICIT' must be in capital
        letters.
     Input format: A8
     Example of input --> EXPLICIT

HTRPOS Immersion heater position.
     Input list: TOP
          or
          SIDE
          or
          NONE
     Note: 'TOP', 'SIDE' or 'NONE' must be in capital
        letters.
     Input format: A4
     Example of input --> NONE

MDIUMF Integer indicator for type of heat exchanger
     fluid.
     Input list: 0 (this sets water as transfer medium)
          or
          1 (this sets water/glykol mixtures as
          transfer medium)
     Input format: free format (integer number)
     Example of input --> 0

MDIUMS   Integer indicator for type of storage material.
                Input list: 0 (this sets water as transfer medium)
                                    or
                            1 (this sets water/glykol mixtures as
                            transfer medium)
                Input format: free format (integer number)
                Example of input --> 0

HTRLEN   Length of storage immersion heater if HTRPOS is
[m]   set to "TOP".
                or
                Distance between top of the store and storage
                immersion heater if HTRPOS is set to "SIDE".
                or
                Any dummy number if HTRPOS is set to "NONE".
                Input format: free format (decimal number)
                Example of input --> 0.46

STOHEI   Height of the store.
[m]   Input format: free format (decimal number)
                Example of input --> 1.117

CONDS   Conductivity of the storage medium.
[W/(mK)]   Input format: free format (decimal number)
                Example of input --> 0.6

VS   Volume of the storage medium.
[litre]   Input format: free format (decimal number)
                Example of input --> 185.0

VF   Volume of the heat exchanger fluid.
[litre]   Input format: free format (decimal number)
                Example of input --> 2.0

QSMDAS   Measured storage capacity rate.
                Note: input 0.0 if it is unknown.
[KJ/K]   Input format: free format (decimal number)
                Example of input --> 781.0

NF   Number of heat exchanger fluid nodes in the model.
                Input format: free format (integer number)
                Example of input --> 1

NS   Number of storage nodes in the model.
                Input format: free format (integer number)
                Example of input --> 8

NDEAD   Number of "dead" nodes in the model.
               Input format: free format (integer number)
               Example of input --> 0

NFSTA   Nodal position of the heat exchanger fluid inlet
               temperature.
               Input format: free format (integer number)
               Example of input --> 1

TFINI   This input (positive only) value will be used as
[deg.C]   the initial condition of the heat exchanger. It is
               assumed that the initial temperature of the heat
               exchanger fluid is uniform and taken equal to
               TFINI.
               Note: negative number will use the default setting
                     (i.e. the first record of the experimental
                     exit temperature from the first data set
                     will be used as the initial condition of the
                     heat exchanger).
               Input format: free format (decimal number)
               Example of input --> -1.0

TSINI   This input (positive only) value will be used as
[deg.C]   the initial condition of the store. It is assumed
               that the initial temperature of the storage
               material is uniform and equal to TSINI.
               Note: negative number will use the default setting
                     (i.e. the first record of the experimental
                     exit temperature from the first data set
                     will be used as the initial condition of the
                     store.)
               Input format: free format (decimal number)
               Example of input --> -1.0

NDATPA   Array for keeping number of experimental data
               points for each mode.
               Note: at least 2 data points for each mode and try
                     not to mix different modes in a set of data.
               Input format: free format (integer number)
               Example of input --> (see below)

UAA   Array for keeping the fluid-to-storage medium heat
[W/K]   transfer coefficient (UAfs) for each mode.
               Input format: free format (decimal number)
               Example of input --> (see below)

UAFAA   Array for keeping the fluid-to-ambient loss
[W/K]   coefficient (UAfa) for each mode.
               Input format: free format (decimal number)
               Example of input --> (see below)

UASAA   Array for keeping the storage medium-to-ambient
[W/K]   loss coefficient (UAsa) for each mode.
               Input format: free format (decimal number)
               Example of input --> (see below)

UASTJA   Array for keeping the storage medium-to-medium
[W/K]   nodal turbulent diffusion coefficient (UAstj) for
               each mode.
               Input format: free format (decimal number)
               Example of input --> (see below)

EXAMPLE 1: The first set of 2 data points is recorded for
            stabilization phase using the discharge loop (i.e.
            IST = -1). The second set of 2242 data points is
            for step charge phase (i.e. IST = 1).

NDATPA	UAA	UAFAA	UASAA	UASTJA
2	280.0	0.0	3.21	0.0
2242	280.0	0.0	3.21	0.0

EXAMPLE 2: The first set of 20 data points is a stabilization
            phase using direct discharge loop (i.e. IST = -1),
            the second set of 960 data points is a half and
            full power charge phase (i.e. IST = 1), the third
            set of 1380 data points is a discharge phase (i.e.
            IST = -1), the fourth set of 960 data points is a
            full and half power charge phase (i.e. IST = 1),
            the fifth set of 360 data points is a discharge
            phase (i.e. IST = -1), the sixth set of 72 data
            points is a stand-by phase (i.e. IST = 0), the
            seventh set of 1262 data points is a complete
            discharge phase (i.e. IST = -1).

NDATPA	UAA	UAFAA	UASAA	UASTJA
20	150.0	1.9	2.0	0.0
960	300.0	1.9	2.0	0.0
1380	150.0	1.9	2.0	0.0
960	300.0	1.9	2.0	0.0
360	150.0	1.9	2.0	0.0
72	250.0	1.2	1.7	0.0
1262	150.0	1.9	2.0	0.0

4. THE PARAMETER IDENTIFICATION USAGE PROGRAM MRQ-4PORT

4.1. Introduction

The parameter identification method MRQREG was developed at the TPD for the evaluation of measurements on passive solar energy systems. One of the aims, however, was to obtain a tool which could be used for a wider range of applications.

In the course of the SSTG project the usefulness of a parameter identification technique was proven, in particular by the activities of the Ecole Nationale Superieure des Mines in this area (see [2], Part B Chapter 2).

Since MRQREG was available in a version which could run on a IBM AT^R compatible machine, it was decided to adapt MRQREG for the evaluation of storage tests.

The main adaptation was the linking of the storage calculation model 4PORT to the identification routines; this combination is called MRQ-4PORT.

Since MRQ-4PORT was born in a rather late stage of the project, the experiences are based on only a limited number of runs.

Moreover, the method MRQREG is a powerful tool; the combination with 4PORT has not been protected against misuse.

Finally, the version of 4PORT in the tool MRQ-4PORT has not in every detail the same possibilities as the versions in DYNAMI and IDENUA, since the latest version of the latter were completed just shortly before the end of the research contract. Possibilities not yet introduced into the available version of MRQ-4PORT are:

- embedded electrical auxiliary heater;
- dead zone in heat exchanger;
- temperature dependent heat exchanger overall coefficient of heat transfer, $(UA)_{f,s}$, .

The main application area for the available version is for fitting alone or in combination parameters like:

- the $(UA)_{s,s}$-value between successive storage segments (see part B chapter 2);
- the effective storage and/or heat exchanger volume;

- the (constant) $(UA)_{f,s}$-value of a heat exchanger, e.g. during a charge
  step test;
- the heat loss overall coefficient of heat transfer $(UA)_{s,a}$.

4.2. Parameter identification

4.2.1. Principle

The principle of parameter identification is illustrated in figure 4.1:
- the thermal system is represented by a mathematical model;
- the values for the parameters in the model are the unknowns, which are to
  be identified by an iterative process of comparing the measured system
  response with the response calculated by the model for the same
  conditions and same time sequence of input data, with assumed parameter
  values;
- the process of iteration continues until an optimal agreement is reached.

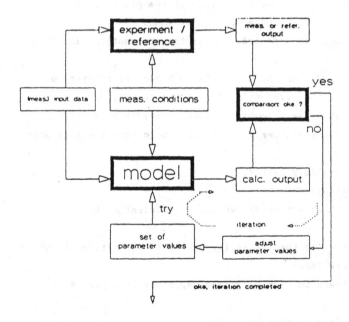

Figure 4.1: The principle of parameter identification.

4.2.2. The method MRQREG

The method MRQREG uses the Levenberg-Marquardt procedure:
the parameter values are adjusted on the basis of the square of the
deviation between measured (the "reference") and calculated response and
the partial derivatives of the response to each of the parameters.
A detailed description of the principle can be found in [2], Part B
Chapter 2.

The main specific feature of MRQREG is, that the partial derivatives are
computed numerically, by applying a small change in parameter value (ΔA)
and looking at the influence on the response (ΔY); so $dY/dA(k)$ is
approximated by $\Delta Y/\Delta A(k)$.

The obvious disadvantage of this approach is the increased computation time
involved: for each parameter k to be fitted an extra model run is needed to
provide $\Delta Y/\Delta A(k)$, at each iteration step.

The advantages are, however, to be found in the good convergence properties
of the Marquardt procedure and in the flexibility to use any kind of model.
The model may be non-linear or otherwise complex. The only and trivial
requirement is that the model gives a response to the input data and the
parameter values and the response is a continuous function of the
parameters to be identified.

In MRQREG, moreover, the algorithms are organized in such a way that the
model calculates a whole time sequence of datapoints in one call without
being interrupted by the identification process. Therefore, the model can
be any existing calculation model with only a few adaptations needed for
creating the link with (becoming a subroutine for) MRQREG.

In MRQREG it is furthermore possible to place a window over the reference
dataset. The calculation will be carried out over the full dataset, but the
fitting procedure will only be carried out over a selected part ("window")
of the whole sequence of datapoints. This makes it possible for the model
to initialize over the minimum number of timesteps needed to "forget" the
influence of the unknown conditions from the period before the start of the
dataset. It also makes it possible to focus on the specific part of the
dataset which is most relevant for the (set of) parameter(s) to be
identified.

Some of the routines in MRQREG are based on ideas from Press et al.
(Numerical Recipes [1]). That publication contains valuable background
information on the subject and a package of standard software routines.

4.3. Description of MRQ-4PORT

4.3.1. General

MRQ-4PORT uses the exit temperature as output variable for the comparison
(see figure 4.1). The input variables are the quantities measured during
the experiment.

MRQ-4PORT uses the following files for input:

MRQ.IN   containing instructions for MRQ actions, including initial values
           for the parameters which are candidates for the identification;

MRQ.DAT   containing the dataset with measured input and output variables;

P4.DAT   containing the configuration of the storage model and the
           dimensions and properties of the modelled storage system,
           excluding the parameters which are candidates for the
           identification.

The following files are generated as output:

MRQ.OUT   the result of the fitting operation;

MRQ.LOG   a log file of the fitting process;

MRQ.PLT   a plot file with function output and input.

4.3.2. Controlling MRQREG via MRQ.IN

Example of MRQ.IN file:

```
IFORM    KFREZE    MFIT    MA
 0         0        2      7
LISTA(1-MFIT)
  1    6
UA     UASA   UAFA      UASKJ   UASTJ   VF     VS
200.   2.5    0.0       0.0     00.0   30.0   200.0    0.0   0.0   0.0
 IWOPEN    IWCLOS
  120        1300
HEADER LINE FOR OUTPUT
 NR   TIME(min) Texit  Texit/calc     TI
```

Explanation:

- IFORM controls the reading of the data files with measured data:

        IFORM = 0 : an original STEP data file is used.

        IFORM = 1 : the STEP data file is a converted file with the column
                  time changed into time steps.

- KFREZE controls the fit process:

        KFREZE = 0 : fit, the fitting process starts with the parameter
                  values A $(1 \rightarrow MA)$, MA = total number of parameters,
                  given in MRQ.IN.

        KFREZE = 1 : no fit, run model with the parameter values given in
                  MRQ.IN.

- MFIT represents the number of parameters to be fitted:

        MFIT   = 0 : (KFREZE = 1), no fit.

        MFIT   = # : (KFREZE = 0), total number of model parameters to be
                fitted.

- MA represents the <u>total</u> number of model parameters given in MRQ.IN.

- LISTA is the list of the model parameter array elements that are to be
        fitted.

        Example: if LISTA $(1 \rightarrow MFIT)$ = 1, then only the UA-value is fitted.
               (MFIT = 1).
               If LISTA $(1 \rightarrow MFIT)$ = 1 2 6, then UA, UASA and VF are
               fitted. (MFIT = 3).

- $A(1 \rightarrow MAMX)$ are the start values of the model parameters. MAMX = 7.
               If less than 7 model parameters are applied, add 7 - MA
               dummy values.

UA   is the UA-value of the heat transfer from fluid to storage
        material given in W/K.

UASA   is the heat loss of the storage material (W/K).

UAFA   is the heat loss of the fluid material (W/K).

UASkj   is the heat conduction between storage nodes due to conduction
        (W/K).

UAStj   is the heat conduction between storage nodes due to turbulances
        and mixing (also: only during the storage flow mode - discharge)
        (W/K).

VF   is the fluid volume (1).

VS   is the storage volume (1).

- IWOPEN is the start value of the window used for fitting.

- IWCLOS is the end value of the window. If IWCLOS - 0, then IWCLOS is set
         to NDATA (which is the number of records of the data file).

- "header line for output" is printed as first line in MRQ.PLT to clarify
  the printed matter.

The data are separated by spaces. The program reads the data unformatted.
Each record of data is preceded by a record with information concerning
these data.

4.3.3. Data file MRQ.DAT

The data file MRQ.DAT should be an original STEP data file (IFORM - 0, see
MRQ.IN) or a data file with an equal STEP format but with records beginning
with time step instead of the time (IFORM - 1).

Format for the original STEP data files, PAUX is not added:

    (I4, 5I2, 2F6.2, F7.3, F7.1, 2F6.2, I3) if IFORM - 0

Format for the converted data files:

    (F14.2, 2F6.2, F7.3, F7.1, 2F6.2, F3.0)

4.3.4. <u>Data file P4.DAT</u>

The data file P4.DAT contains the following model parameters:

NF   : number of fluid nodes
NS   : number of storage nodes
NDEAD   : number of nodes in the dead zone
IYFUC   : number of the fluid segment used for recording the model
                  (charge) exit temperature
IYFUD   : number of the storage segment used for recording the model
                  (discharge) exit temperature
TFM (i=1,NF)   : initial fluid temperatures
TSM (i=1,NS)   : initial storage temperatures
CPFLU   : specific heat of the fluid segments J/kg/K
CPSTO   : specific heat of the storage segments J/kg/K
RHOFLU   : specific mass of the fluid segments kg/m³
RHOSTO   : specific mass of the storage segments kg/m³
CFLOW   : conversion factor for the flow rate. The recorded flow rate
                  multiplied with CFLOW should have the dimension of m³/h.
                  This means that if the recorded flow rate has the dimension
                  of 1/h CFLOW should be 0.001.

Example of a P4.DAT data file:

```
NF   NS   NDEAD   IYFUC   IYFUD
6    10     0       6       1
TFM(..............)
19.94  19.94  19.94  19.94  19.94  19.94
TSM(..............)
19.94  19.94  19.94  19.94  19.94  19.94  19.94  19.94  19.94  19.94
CPFLU  CPSTO  RHOFLU  RHOSTO
4185.  4185.   998.    998.
CFLOW
 0.001
```

The data are separated by spaces. The program reads the data unformatted.
Each record of data is preceded by a record with information concerning
these data.

4.3.5. Output file MRQ.OUT

Example of a MRQ.OUT output file:


```
******** Calculated with program MRQREG  ********

Report :
          Date : 00-00- 00           Time : 00:00

          Simulation with model 4-Port
          window:
             this window is used for the
             fit procedure.

             first number I for fit procedure  :   120
             last  number I for fit procedure  :  1300

             total number of data records used
             for the final calculation of the
             standard deviation.                     1217
          Number of iterations (if applicable):      4

Coefficients and covariance :

k     Coeff.    Covar.      Covar.      Covar.      Covar.      Covar.      Covar.      Covar.
1   283.        27.0       0.000E+00   0.000E+00   0.000E+00   0.000E+00   -4.09       0.000E+0
2   2.50       0.000E+00   0.000E+00   0.000E+00   0.000E+00   0.000E+00   0.000E+00   0.000E+0
3   0.000E+00  0.000E+00   0.000E+00   0.000E+00   0.000E+00   0.000E+00   0.000E+00   0.000E+0
4   0.000E+00  0.000E+00   0.000E+00   0.000E+00   0.000E+00   0.000E+00   0.000E+00   0.000E+0
5   0.000E+00  0.000E+00   0.000E+00   0.000E+00   0.000E+00   0.000E+00   0.000E+00   0.000E+0
6   40.5       -4.09       0.000E+00   0.000E+00   0.000E+00   0.000E+00   0.844       0.000E+0
7   200.       0.000E+00   0.000E+00   0.000E+00   0.000E+00   0.000E+00   0.000E+00   0.000E+0

Chisq over window : 110.59921     Sdev :   0.30628

Chisq over dataset: 221.48923     Sdev :   0.42679
```


Explanation:

Coefficients : the parameters $A(k)$;

Covariance   : the covariance matrix for the identified parameters;
               if covar $(k,1) = 0$: the parameter $A(k)$ has been kept frozen
               at its initial value (see MRQ.IN).

Chisq   : the sum of the square of deviation between measured and
               calculated exit temperature;

Sdev   : the standard deviation between measured and calculated exit
               temperature (in K).

4.3.6. <u>Output file MRQ.PLT</u>

The output file MRQ.PLT contains 5 columns with the following data:

NR   number of the data record;

TIME(MIN)   time (minutes);

Texit   exit temperature, measured (°C);

Texit/calc exit temperature, calculated (°C);

TI   inlet temperature, measured (°C).

REFERENCES

[1] Press, W.H., et al.,
     Numerical Recipes, The Art of scientific computing,
     Cambridge University Press, Cambridge, 1987;
     ISBN 0521 30811 9

[2] Visser, H. and H.A.L. van Dijk (ed.)
     Final Report on the Activities of the Solar Storage Testing Group.
     Commission of the European Communities, EUR 13119.

APPENDIX E:

RECOMMENDATIONS FOR WRITING THE TEST REPORT

<u>CONTENTS</u>

1. <u>INTRODUCTION</u>

The test report of a heat storage system should contain the following:
- general description of the storage system
- thermal characterization of the storage material
- test facility description, and
- thermal characterization of the storage system.
All these items are discussed briefly in the following chapters.

2. GENERAL DESCRIPTION OF THE STORAGE SYSTEM

The heat storage device should be defined by its name and the name and
address of the manufacturer. The commercial status of the store (i.e.
whether it is commercially available or a prototype) should be indicated.
Further, the area of use to which the store will be put is also important.
In the general description of the storage system the main storage material
and its method of encapsulation should be described briefly together with
the physical properties of the heat transfer fluid. Recommendations for a
detailed characterization of the storage material are given in Chapter 3.
The geometry of the entire storage device should be given and the required
orientation should be indicated. Special (external) components should be
described.

3. THERMAL CHARACTERIZATION OF THE STORAGE MATERIAL

The characterization of the storage material is particularly important for latent heat storage devices. The commercial name of the storage material and the name and address of the manufacturer should be given. The main constituents should be indicated by their chemical formulae, weight and volume percentages. Also, additives such as nucleating agents, stabilizers and/or additives to enhance the thermal conductivity should be described. The heat storage material should be characterized by the following properties, bearing in mind both heating and by cooling situations:

- the apparant specific heat capacity as a function of temperature
- the transition temperature (range)
- the latent heat, and
- the mean specific heat, density and effective thermal conductivity (including natural convection) below and above the transition temperature (range).

Ageing effects should be described. These should be based on the transition temperature (range) and latent heat as a function of the number of cycles.

4. TEST FACILITY DESCRIPTION

The test facility may be described in the test report or reference may be
made to a separate test facility description and calibration report.
Each heating and cooling loop of the test facility should be described. A
general explanation should be given of the maximum heating or cooling
power, the minimum and maximum heat transfer fluid flow rate, the pipework
and fittings, the pump as well as concerning the control of the ambient air
temperature, the flow rate, the storage inlet temperature and the heating
power. The quality of the temperature step and the short and long term
stability of the test loops may be elucidated as well.
The test facility description should also contain information on the
equipment used for measuring the flow rate, storage inlet and/or outlet
temperature, storage differential temperature and ambient temperature. The
type of sensor used for the measurement, the location of the sensor in the
test facility or test area and the calibration data should be given.
Additional measuring equipment should be reported too.
A description of the calibration procedure should be given. The calibration
result should be the overall accuracy (i.e. the accuracy of the final
output after passage through the chain of instruments) of the quantity
measured by the sensor. If possible, a distinction should be made between
the systematic and random part of the error. The systematic errors should
be corrected as much as possible. For instance, the differential
temperature sensor should be corrected for offset and heat loss (see
Appendix A). The size of random errors should be presented as two times the
standard deviation (see Appendix B). If it is not possible for a
distinction between the systematic and random part of the error to be made
(or, at least, estimated), then the whole error should be considered to be
systematic.

5. THERMAL CHARACTERIZATION OF THE STORAGE SYSTEM

The thermal characterization of the heat storage device is the most important part of the test report. It describes the results of the heat storage test procedures.

At the very least, the following quantities of the storage system tested should be reported:
- the overall measured and theoretically calculated coefficient of heat transfer between the store and the ambient environment;
- the heat storage capacity;
- the heat storage efficiency at different times; and
- (if a heat exchanger is present) the overall coefficient of heat transfer between transfer fluid and storage material and the temperature and flow rate measuring conditions.

If a more detailed characterization of a heat storage is desirable or necessary (for criteria see Part B Chapter 7) one or more of the following quantities should also be reported:
- the overall coefficient of heat transfer between store and ambient environment during stand-by;
- the overall coefficients of heat transfer between specific parts of the store and the ambient environment;
- the effective thermal conductivity within the store under flow conditions and/or during stand-by;
- the overall coefficient of heat transfer between the auxiliary heater transfer fluid and the storage material.

All the tests, test conditions and the data processing methods used for the determination of the quantities should be described briefly.

The results from the dynamic test should also be given.

For reporting the thermal characterization of storage systems the output of the data processing programs such as STEP, IDENUA, MRQ-4PORT and DYNAMI may be of use.

For the - optional - presentation of the inlet and outlet temperature as a function of time from a temperature step response test, it may be convenient to use the following non-dimensional quantities:
- for storage outlet temperature, θ_e;

- for storage inlet temperature, θ_i;
- for time, τ.

Definitions can be found in the Nomenclature section at the beginning of Part B.